PERGAMON INTERNATIONAL LIBRARY
of Science, Technology, Engineering and Social Studies

The 1000-volume original paperback library in aid of education, industrial training and the enjoyment of leisure

Publisher: Robert Maxwell, M.C.

LIPIDS IN FOODS
CHEMISTRY, BIOCHEMISTRY AND TECHNOLOGY

Pergamon publications of related interest

BOOKS

BIRCH, CAMERON & SPENCER
Food Science, 2nd Edition

CHRISTIE
Lipid Analysis, 2nd Edition

CROMPTON
Additive Migration from Plastics into Food

ERICKSSON
Maillard Reactions in Food

GALLI & AVOGARO
Polyunsaturated Fatty Acids in Nutrition

GAMAN & SHERRINGTON
The Science of Food, 2nd Edition

JOURNALS

PROGRESS IN FOOD AND NUTRITION SCIENCE

PROGRESS IN LIPID RESEARCH

LIPIDS IN FOODS CHEMISTRY, BIOCHEMISTRY AND TECHNOLOGY

by

FRANK D. GUNSTONE

The University of St. Andrews, Scotland

and

FRANK A. NORRIS

Glenview, Illinois, U.S.A.

PERGAMON PRESS

OXFORD · NEW YORK · TORONTO · SYDNEY · PARIS · FRANKFURT

U.K.	Pergamon Press Ltd., Headington Hill Hall, Oxford OX3 0BW, England
U.S.A.	Pergamon Press Inc., Maxwell House, Fairview Park, Elmsford, New York 10523, U.S.A.
CANADA	Pergamon Press Canada Ltd., Suite 104, 150 Consumers Rd. Willowdale, Ontario M2J 1P9, Canada
AUSTRALIA	Pergamon Press (Aust.) Pty. Ltd., P.O. Box 544, Potts Point, N.S.W. 2011, Australia
FRANCE	Pergamon Press SARL, 24 rue des Ecoles, 75240 Paris, Cedex 05, France
FEDERAL REPUBLIC OF GERMANY	Pergamon Press GmbH. 6242 Kronberg-Taunus, Hammerweg 6, Federal Republic of Germany

First edition 1983

Library of Congress Cataloging in Publication Data

Gunstone, F. D.
Lipids in foods.

(Pergamon international library of science, technology, engineering, and social studies)
1. Lipids. 2. Food industry and trade. I. Norris, Frank A. II. Title. III. Series.
TP453.L56G86 1983 664′.3 82-16498

British Library Cataloguing in Publication Data

Gunstone, Frank D.
Lipids in foods.—(Pergamon international library)
1. Food 2. Acids, Fatty
I. Title II. Norris, Frank A.
664′.3 QD305.A2

ISBN 0-08-025499-3
ISBN 0-08-025498-5 Pbk

Printed in Great Britain by A. Wheaton & Co. Ltd., Exeter

Preface

Natural and processed fats and oils, whether of animal or vegetable origin, play a significant role in the economy of several countries including both oil-producers and oil-users. These materials are used extensively, but not exclusively, in the food industry and an increasing number of students, researchers, technologists, and industrialists require basic information on the chemistry, biochemistry and technology of the fatty acids that we now call lipids. This book, which required two authors, seeks to meet this need.

One of us (Professor Frank Gunstone), writing from an academic and research background with considerable experience in teaching in this field, has contributed the first ten chapters covering the basic chemistry and biochemistry of the fatty acids and their natural derivatives. This includes an account of the chemical structure, separation, analysis, biochemistry, physical properties, chemical properties, and synthesis of these compounds.

The remaining chapters are written by Dr. Frank Norris, formerly Senior Group Leader, Edible Oil Products, Kraft R & D, Glenview, Illinois, and formerly Head, Oil Mill and Refinery Research, Swift & Company, R & D, Chicago, Illinois, U.S.A. The topics covered include the recovery of fats and oils from their sources and the processes of refining, bleaching, hydrogenation, deodorisation, fractionating, and interesterification. The final chapters are devoted to margarines and shortenings and to the problems of flavour stability and antioxidants.

We believe that the value of our book is enhanced by including both science and technology, so that students wishing to know more about lipids (and those already working with them) can learn how they are made in nature and how they are recovered, purified, and modified by man for his own needs.

St. Andrews — F. D. GUNSTONE
Glenview — F. A. NORRIS

February 1982

Contents

CHAPTER 1

The Structure of Fatty Acids and Lipids

A. FATTY ACIDS

1. Fatty acid nomenclature

The terms fatty acid and long-chain acid are used interchangeably in this book to describe any aliphatic acid, usually with a chain of ten or more carbon atoms, which occurs naturally in fats, oils, and related compounds (lipids) and also some other acids of closely related structure. Trivial names which were often given to long-chain acids before their chemical structure was fully elucidated, frequently refer to the natural source or to the physical appearance of the acid in question. Thus lauric acid was first isolated from seed fats of the Lauraceae, myristic acid from seed fats of the Myristiceae, and palmitic acid from seed fats of the Palmae. The name margaric acid (given to what was wrongly thought to be the C_{17} alkanoic acid) relates to the pearl-like crystalline form of this substance (Gk *margaron* pearl). Many of these trivial names are so fully embedded in chemical and technical literature that their use can hardly be avoided.

Systematic names, based on IUPAC nomenclature, indicate the chain-length of the acid, the position, nature, and configuration of any unsaturated centres, and the position and nature of any substituents. Two examples are detailed below (**2** and **3**).

A useful shorthand system uses a series of numbers and symbols to indicate the structure of long-chain acids. The symbol 18:1 indicates an unbranched C_{18} acid with one unsaturated centre. The more extended representations 18:1 9*c* or 18:1 9*Z* show the presence of a double bond of *cis* or *Z* configuration starting on the ninth carbon atom (COOH=1). The symbols *Z* and *E* represent the two possible configurations of a pair of geometric isomers. In most fatty acids the *Z* and *E* forms correspond to the *cis* and *trans* isomers respectively. In these formulations the letters *c* (or *Z*), *t* (or *E*), *e*, or *a* indicate *cis* olefinic unsaturation, *trans* olefinic unsaturation, olefinic unsaturation of unknown configuration or where there is no isomerism, and acetylenic unsaturation respectively.

In IUPAC nomenclature olefinic centres are numbered according to their relation to the carboxyl group so that acid (**1**) is octadec-6-enoic acid* indicating that unsaturation is between C-6

$$CH_3(CH_2)_{10}CH{=\!=}CH(CH_2)_4COOH \qquad (1)$$

and C-7. In special circumstances to be detailed later it is useful to count from the methyl (CH_3) end of the chain. The double bond would then be indicated by the symbol *n*-12 (formerly represented as ω12). Since natural polyene acids usually display a common pattern of methylene-interrupted unsaturation, the designation of one double bond also signifies the remainder. This system is frequently used throughout this chapter.

These points are illustrated in the following examples of linoleic (**2**) and arachidonic acid (**3**):

$$CH_3(CH_2)_4CH \overset{cis}{=} CHCH_2CH \overset{cis}{=} CH(CH_2)_7COOH$$

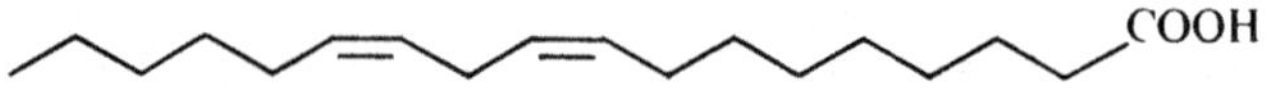

18:2 9*c*12*c* or 18:2 9*Z*12*Z* or 18:2 *n* - 6
octadeca - 9*c*,12*t* - dienoic acid*
linoleic acid (**2**)

$$CH_2(CH_2)_4CH{=}CHCH_2CH{=}CHCH_2CH{=}CHCH_2CH{=}CH(CH_2)_3COOH$$

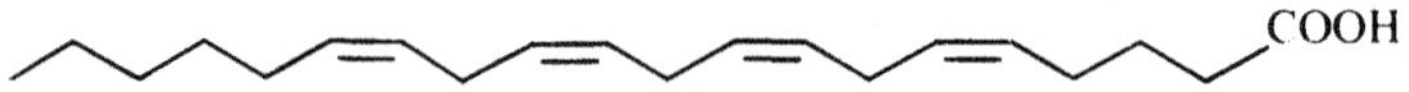

20:4 5*c*8*c*11*c*14*c* or 20:4 5*Z*8*Z*11*Z*14*Z* or 20:4 *n*– 6
icosa - 5*c*, 8*c*, 11*c*, 14*c* - tetraenoic acid*
arachidonic acid (**3**)

2. General comments on structure and occurrence

The number of natural long-chain acids occurring in lipids probably exceeds 800. Most of these are rare and of limited importance and although only the more common acids are of interest for the purposes of this book, the structures of the larger number of natural acids along with a general idea of their frequency of occurrence makes it possible to make generalisations about the structure of natural fatty acids. The following statements have general validity even though there are exceptions to each of them. Each is a consequence of the biosynthetic pathways outlined in Chapter 4.

(i) Most natural acids, whether saturated or unsaturated, are straight-chain compounds with an even number of carbon atoms in each molecule (odd acids, branched-chain acids, and cyclic acids also exist).

(ii) Although the range of chain-length is great (C_2 to $>C_{80}$), the most common chain-lengths are C_{16}, C_{18}, C_{20}, and C_{22}.

(iii) Monounsaturated acids usually contain a *cis*(*Z*) olefinic bond present in one of a limited number of preferred positions (*trans* (*E*) alkenoic and alkynoic acids are also known).

(iv) Most polyunsaturated acids have two to six *cis*(*Z*) double bonds arranged in a methylene-interrupted pattern as in linoleic (**2**) and arachidonic acid (**3**) (acids with conjugated unsaturation and other polyene arrangements also exist).

(v) Substituted acids are uncommon but hydroxy, oxo, and epoxy acids are known. Only branched-chain, cyclic, and substituted acids are potentially enantiomeric (optically active).

Based on the 1964–1970 production of commercial oilseeds it has been estimated that the eight acids listed in Table 1.1 account for 97% of all those produced by plants. The major acids in animal

* The position of the number presents some difficulties. They are generally placed as shown here by British authors. American authors put them first as in 9*c*,12*c*-octadecadienoic acid.

fats and fish oils are palmitic (16:1), palmitoleic (16:1), oleic (18:1), arachidonic (20:4 *n*-6), icosapentaenoic (20:5 *n*-3), and docosahexaenoic acids (22:6 *n*-3).

TABLE 1.1. *The More Common Acids produced by Plants*

Trivial name	Symbol	%
lauric	12:0	4
myristic	14:0	2
palmitic	16:0	11
stearic	18:0	4
oleic	18:1	34
linoleic	18:2	34
α-linolenic	18:3	5
erucic	22:1	3

3. Saturated (alkanoic) acids

The more important saturated acids are listed in Table 1.2 along with some physical properties of the acids and their methyl esters.

TABLE 1.2. n-*Saturated Acids and Their Methyl Esters*

Symbol	Systematic name	Trivial name	Acid mp*	Acid bp*	Ester mp*	Ester bp*
12:0	dodecanoic	lauric	44.8	130^1	5.1	262
14:0	tetradecanoic	myristic	54.4	149^1	19.1	114^1
16:0	hexadecanoic	palmitic	62.9	167^1	30.7	136^1
18:0	octadecanoic	stearic	70.1	184^1	37.8	156^1
20:0	icosanoic	arachidic	76.1	204^1	46.4	188^2
22:0	docosanoic	behenic	80.0	—	51.8	206^2
24:0	tetracosanoic	lignoceric	84.2	—	57.4	222^2

* Values in °C; bp at 760 mm Hg or at other pressures indicated by the superscript.

Lauric and myristic acids occur in coconut oil (45–50 % and 15–18 % respectively) and in palm kernel oil (45–55 % and 15–18 % respectively). Palmitic acid is the most common saturated acid in most seed oils (see Table 1.1) and attains useful levels in cottonseed oil (22–28 %) and palm oil (35–40 %). Stearic acid is present in many ruminant depot fats (tallows ~30 %) and in cocoa butter (~35 %). It can be obtained easily by complete hydrogenation of the readily available unsaturated C_{18} acids (see Table 1.1). The higher saturated acids listed in Table 1.2 are less readily available.

It is apparent from the data in Table 1.2 that the saturated acids are solids with melting points which increase with molecular weight. Unsaturated acids, on the other hand, are liquids or low-melting solids. In general, therefore, the melting range of a natural fat depends on the proportion of saturated and unsaturated acids which it contains and to a lesser extent on the chain-length of the acids.

4. Monoenoic (alkenoic) acids

Over 100 monoenoic acids are known, of which oleic acid (**4**) is by far the most common (see Table 1.1).

$$CH_3(CH_2)_7CH \overset{c}{=\!=} CH(CH_2)_7COOH \qquad (4)$$

Some other monoenoic acids are also listed in Table 1.3. Those with *trans*(*E*) configuration occur only rarely in natural sources, but are significant in other connections. Oleic acid is easily isolated from olive oil or almond oil in which it is a major component. Hexadecenoic acid occurs widely in animal fats, but rarely attains a high proportion. Erucic acid is a major component of rape oil (mustard oil, colza oil), though erucic-free varieties are now being grown.

TABLE 1.3. *Monoenoic Acids*

Symbol	Systematic name	Trivial name*	mp °C *cis*	mp °C *trans*
14:1 9	tetradec-9-enoic	myristoleic	−4	18.5
16:1 9	hexadec-9-enoic	palmitoleic	0.5	32
18:1 6	octadec-6-enoic	petroselinic	29	54
18:1 9	octadec-9-enoic	oleic	16	45
18:1 11	octadec-11-enoic	*cis*-vaccenic	15	44
22:1 13	docos-13-enoic	erucic	34	60
24:1 15	tetracos-15-enoic	nervonic	41	65.5

* Name of the *cis* isomer.

5. Methylene-interrupted polyenoic acids

The common and important polyene acids have two to six *cis*(*Z*) double bonds arranged in a methylene-interrupted pattern as already illustrated for linoleic acid (**2**) and arachidonic acid (**3**). Unsaturated centres are separated from each other by a single methylene group. The polyene acids are usually divided into families depending on the relationship between the methyl group and its nearest double bond. The most important groups are listed in Table 1.4. The acids in each family are produced biosynthetically and sequentially from the first member in the list. Several isomeric pairs occur in the lists (e.g. 20:4 *n*-6 and 20:4 *n*-3).

TABLE 1.4. *Methylene-interrupted Polyene Acids*
(all unsaturated centres have *cis*(*Z*) configuration)

n-9 acids	*n*-6 acids	*n*-3 acids
18:1 9	18:2 9, 12[a]	18:3 9, 12, 15[d]
18:2 6, 9	18:3 6, 9, 12[b]	18:4 6, 9, 12, 15
20:2 8, 11	*20:3 8, 11, 14	20:4 8, 11, 14, 17
20:3 5, 8, 11	*20:4 5, 8, 11, 14[c]	*20:5 5, 8, 11, 14, 17[e]
22:3 7, 10, 13	22:4 7, 10, 13, 16	22:5 7, 10, 13, 16, 19
22:4 4, 7, 10, 13	22:5 4, 7, 10, 13, 16	22:6 4, 7, 10, 13, 16, 19[f]

* Precursors of prostaglandins and thromboxanes.
[a]linoleic; [b]γ-linolenic; [c]arachidonic; [d]α-linolenic; [e]icosapentaenoic; [f]docosahexaenoic.

Some of the *n*-6 polyene acids (and possible also of the *n*-3 series) are essential dietary requirements for the healthy growth of animals. The physiological purposes of these acids are probably not fully recognised, but animals raised on a diet lacking these essential fatty acids develop deficiency symptoms which are best cured by arachidonic acid (20:4 *n*-6) but also by linoleic acid (18:2 *n*-6) which was once called vitamin F. Animals cannot make linoleic or α-linolenic acid for themselves, but most animals can convert these to higher members of the *n*-6 and *n*-3 families respectively. An arachidonic acid deficiency can therefore usually be cured by dietary linoleic acid. Three of the C_{20} acids in Table 1.4 are precursors of the prostaglandins, thromboxanes, and leukotrienes.

Linoleic acid is an important component in many seed oils such as soybean (48–58 %), safflower (55–81 %), or sunflower (20–75 %). α-Linolenic acid is the major acid in linseed oil (55–60 %). Arachidonic acid is present in animal phospholipids (e.g. liver lipids), and the 20:5 and 20:6 *n*-3 acids are major acids in many fish oils.

6. Other acids

Though most of the common acids belong to the above groups some acids falling outside these categories deserve comment.

Branched-chain acids with a methyl group on the penultimate (iso acids) or antepenultimate carbon atom (anteiso acids) are trace components of many animal fats. The iso acids (**5**) generally have an even number of carbon atoms per molecule (including the branched methyl group). The anteiso acids (**6**) are usually odd, contain a chiral centre, and belong to the *S*-(+) series.

$$CH_3CH(CH_3)(CH_2)_nCOOH$$

iso acids (**5**)

$$CH_3CH_2CH(CH_3)(CH_2)_nCOOH$$

anteiso acids (**6**)

Acids containing a cyclopropane (**7**) or cyclopropene (**8**) group are known. The former are mainly of bacterial origin and the latter of plant origin. When ingested the cyclopropene acids produce undesirable effects. For example, egg whites become pink. Biochemically, it is known that sterculic acid inhibits the bio-desaturation of stearic to oleic acid.

$$CH_3(CH_2)_5\overbrace{CHCH}^{CH_2}(CH_2)_9COOH$$

lactobacillic acid (**7**)

$$CH_3(CH_2)_7\overbrace{C{=}C}^{CH_2}(CH_2)_7COOH$$

sterculic acid (**8**)

Reference is made to three types of oxygenated acids. The most common hydroxy acid is ricinoleic (**9**) which is the major acid in castor oil; vernolic acid (**10**) is typical of a group of epoxy acids present in several seed oils, and the furan-containing acid (**11**) is one of a family of such acids sometimes found in fish lipids.

$CH_3(CH_2)_5CH(OH)CH_2CH{=}CH(CH_2)_7COOH$

(+) - 12*R* - hydroxyoleic acid
ricinoleic acid **(9)**

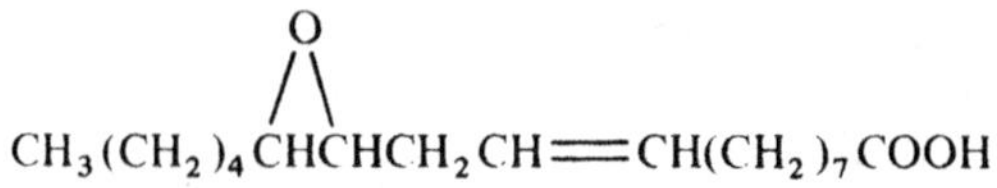

(+) -12*S*, 13*R* - epoxy oleic acid
vernolic acid **(10)**

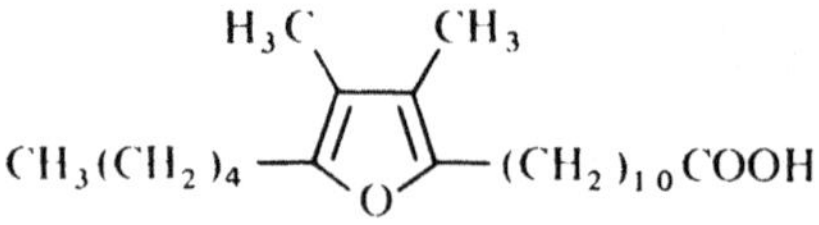

(11)

B. LIPIDS

1. General comments on lipid structure

Long-chain acids of the type which have just been discussed occur in nature in several combined forms known as lipids (Gk *lipos* fat). Most are esters, but some are amides. In addition to the specific names which will be explained in the following sections, a number of general names should be noted. The name glycerolipid applies to any lipid which contains glycerol (propane-1,2,3-triol) among its hydrolysis products; a glycolipid contains a sugar residue, a phospholipid (phosphatide) furnishes phosphoric acid on hydrolysis, a sphingolipid is based on sphingosine or a related long-chain amine, and a sulpholipid contains a sulphate group. All lipids, by definition, also furnish fatty acids and/or related compounds on hydrolysis. Sometimes these terms are combined as in glycosphingolipid, but these are self-explanatory if the simpler terms are understood.

Appropriate derivatives of glycerol contain a chiral centre and may therefore exist in enantiomeric (or racemic) forms which can be described in D/L or *R*/*S* terminology. Both these systems have some disadvantages when applied to lipids and a different system, peculiar to glycerol derivatives, is widely accepted. When glycerol is represented by the Fischer projection **(12)** the three carbon atoms are numbered 1 to 3 from top to bottom and the prefix *sn* (stereospecific numbering) indicates that this system is being used. Structure **(13)** represents an enantiomeric monoacylglycerol. Its enantiomer **(14a)** is 3-acyl-*sn*-glycerol **(14b)**. Similarly, 1,2-diacyl-*sn*-glycerol would be the enantiomer of 2,3-diacyl *sn*-glycerol **(15)**. Most phospholipids are based on the structure **(16)**.

Symbols α and β, sometimes employed to designate groups attached to the primary and secondary glycerol hydroxyl groups respectively, are only appropriate for achiral or racemic compounds.

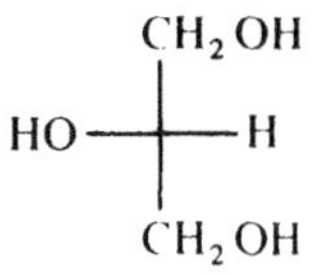

(12)

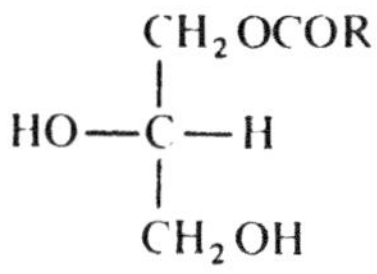

(13)

1-acyl-*sn*-glycerol

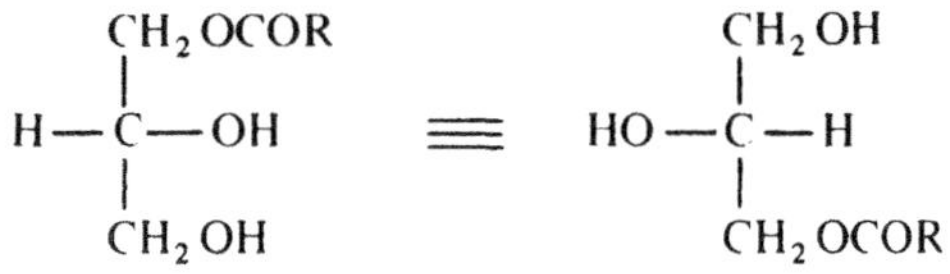

(14a) (14b)

3-acyl-*sn*-glycerol

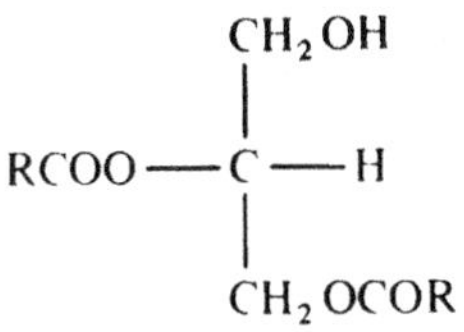

(15)

2, 3-diacyl-*sn*-glycerol

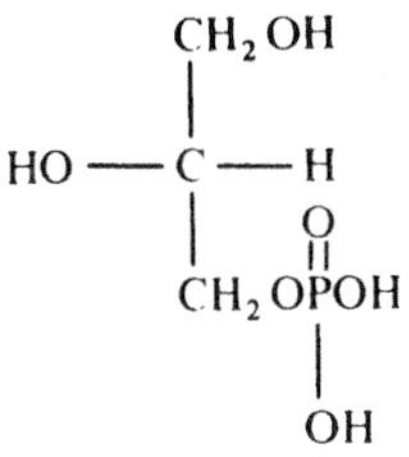

(16)

sn-glycerol 3-phosphate (or, more correctly, *sn*-glycero-3-dihydrogen phosphate)

2. Acylglycerols (glycerides)

The most familiar lipids are acyl esters of glycerol known as monoacylglycerols (monoglycerides), diacylglycerols (diglycerides), and triacylglycerols (triglycerides). The newer terms should be used, but the latter still appear widely in older academic and in technical publications. The triacylglycerols are the most common of these esters and are usually described as fats when solid at ambient temperatures (e.g. mutton fat) and as oils when liquid at ambient temperatures (e.g. soybean oil). Some solid vegetable fats are also called "butters" (e.g. cocoa butter).

Monoacylglycerols exist as 1-acyl (**17**) or 2-acyl (**18**) isomers which may also be designated as α- and β-monoglycerides respectively. The 1- and 3-acyl *sn* isomers are enantiomers which together comprise the racemic α-monoglyceride.

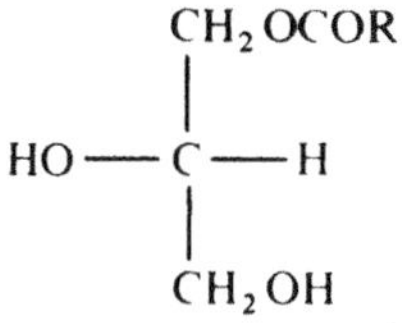

1-acylglycerol (17)

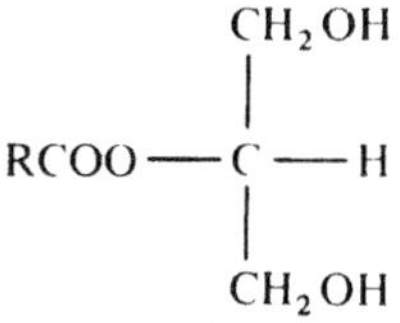

2-acylglycerol (18)

Diacylglycerols exist in $\alpha\beta$ (1,2- and 2,3-diacyl esters, **19**) and $\alpha\alpha'$ (1,3-diacyl esters, **20**) forms.

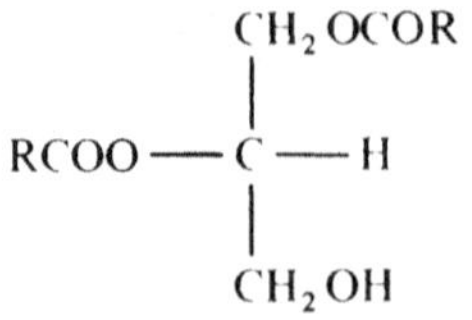

$\alpha\beta$ - diglyceride (19)

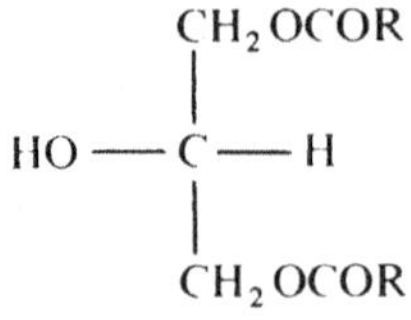

$\alpha\alpha'$ - diglyceride (20)

Natural oils and fats are mainly triacylglycerols. This is the most significant class of storage lipids in plants and in most animals—though not necessarily in marine lipids. Natural triacylglycerols seldom contain three identical acyl groups: more usually they contain two or three different acyl groups selected from all those present in the source. The number of possible triacylglycerols rises quickly with the number of available acyl groups. This point is discussed further in Chapter 3, but it can be noted here that with only two different fatty acids there are eight possible triacylglycerols of which four are achiral and the remainder represent two enantiomeric pairs. With three different acyl groups there are twenty-seven different compounds including nine enantiomeric pairs.

Hydrolysis of acylglycerols by acid or alkali gives glycerol and a mixture of fatty acids. Enzymic hydrolysis is more specific: pancreatic lipase, for example, hydrolyses the acyl groups attached to the two primary hydroxyl groups and leaves the 2-monoacylglycerols (see Chapter 3).

3. Acyl derivatives of alcohols other than glycerol

This is a less common type of lipid, but reference must be made to wax esters and sterol esters.

Waxes are mixtures of compounds which give the surfaces they cover a characteristic sheen and the ability to repel water. This latter property is important in conserving water within the organism and in providing a barrier against the environment. One component of waxes are the wax esters. These are esters of long-chain acids with long-chain hydroxy compounds which may be mono- or dihydroxyalkanes or hydroxy acids. Carnauba wax, a typical vegetable wax, contains about 36% of wax esters. These include C_{46}–C_{64} compounds, but are mainly C_{54} (13%), C_{56} (27%), C_{58} (16%), and C_{60} (17%) esters derived from C_{30} (16%), C_{32} (60%), and C_{34} (17%) alcohols and appropriate fatty acids. Wool wax and beeswax are important animal waxes.

Most samples of extracted lipid contain some free sterol. This is usually cholesterol from animal sources and stigmasterol, β-sitosterol, or ergosterol from plant sources. In addition, some of the sterol is esterified with a fatty acid.

4. Glycosyldiacylglycerols

In these glycolipids glycerol is acylated at the C-1 and C-2 positions and linked glycosidically at C-3 to a mono-, di-, or trisaccharide. In higher plants and algae, and particularly in the chloroplast, the sugar residue is gal(galactose) (**21**), gal-gal, or 6-sulphoquinovose (**22**). The associated fatty acids are highly unsaturated.

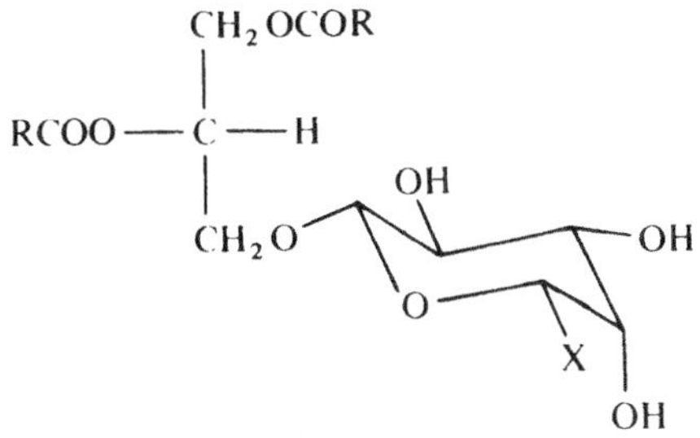

(21) X = CH_2OH monogalactosyl-diacylglycerol (MGDG)

(22) X = CH_2SO_3H plantsulpholipid (sulphoquino vosyldiacylglyceride)

In bacteria, branched-chain fatty acids become significant and galactose may be accompanied or replaced by glucose or mannose or di- and trisaccharides built up from these.

5. **Phosphoglycerides (phosphatidic acids, phosphatidylglycerols, phosphatidylcholine, phosphatidylethanolamine, phosphatidylserine, phosphatidylinositide)**

Phosphoglycerides are derivatives of glycerophosphoric acid or more strictly *sn*-glycerol 3-phosphate (**23**). They occur widely throughout the animal and vegetable kingdom, being particularly associated with biological membranes (Chapter 4). These provide a physical barrier separating cells and subcellular organelles from their environment and are intimately involved in essential life processes. They facilitate and control the transport of metabolites between the environments they separate and are involved in many cell functions.

Most extracted lipid samples contain some phospholipids. These are largely removed from vegetable oils during processing (Chapter 12), but they are ingested when meat, vegetables, or cereals are eaten. It is important to recognise that natural phosphoglycerides, in which the hydroxyls are acylated, represent classes of compounds differing in their acyl moieties rather than single compounds.

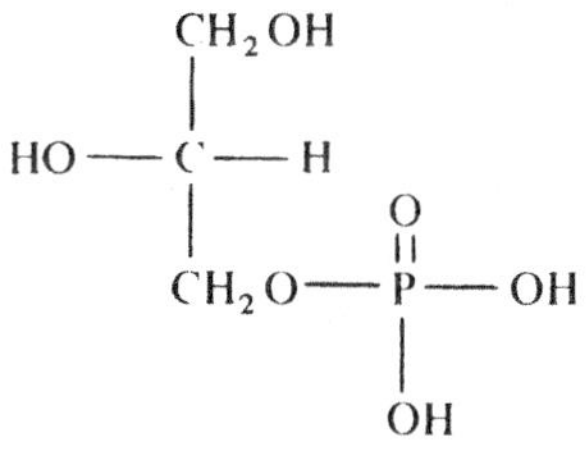

sn-glycerol 3-phosphate (23)

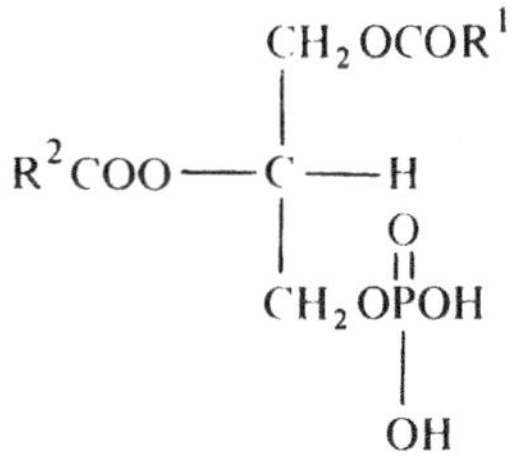

1, 2-diacylglycerol 3-phosphate, phosphatidic acid (PA) (24)

The **phosphatidic acids** (PA, **24**) are derivatives of *sn*-glycerol 3-phosphate with the two glycerol hydroxyl groups acylated by long-chain acids—usually by two different acids. The phosphatidic acids represent the common unit in the phosphoglycerides still to be discussed and are important intermediates in the biosynthesis of these compounds. The ester linkage between glycerol and phosphoric acid is more resistant to hydrolysis—especially by alkali—than the glycerol–fatty acid linkages, so that mild hydrolysis gives fatty acid and glycerol phosphate. The latter is a mixture of α and β isomers because the phosphoric acid unit can move from one hydroxyl to another. Phosphatidic acids are mono esters of the tribasic phosphoric acid which can form further ester linkages. When so combined the phosphatidic acid residue is designated by the term **phosphatidyl**. This includes the two acyl groups attached to the glycerol unit.

Complete hydrolysis of the phosphatidic acids gives long-chain acids, phosphoric acid, and glycerol. Other lipids which furnish the same hydrolysis products, though not in the same proportions, include the **phosphatidylglycerols** (PG, **25**) and the **diphosphatidylglycerols** (cardiolipin, **26**). Notice that these names are single words and are not broken into two. Phosphatidylglycerol is an important component of photosynthetic tissue where it is particularly associated with an uncommon C_{16} acid—hexadec-3*t*-enoic acid. Diphosphatidylglycerol is a major lipid component of mitochondria and when derived from this source it is unusually rich in linoleic acid. Both compounds exist in partially acylated forms. [The use of RCO as the acyl-group in structures **25** and **26** does not imply that all these are the same].

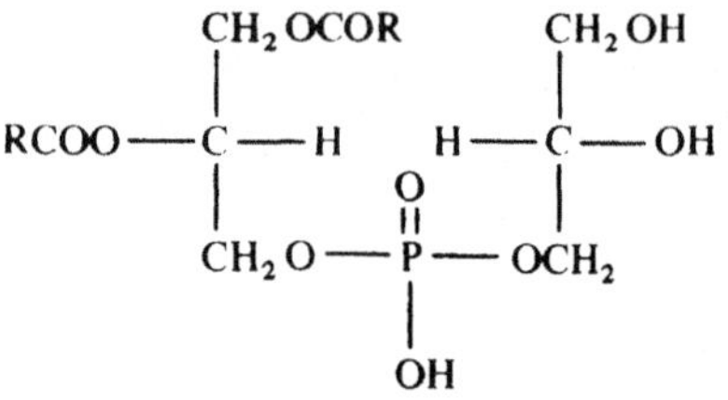

3 - *sn* - phosphatidyl - 1′ - *sn* - glycerol

PG (25)

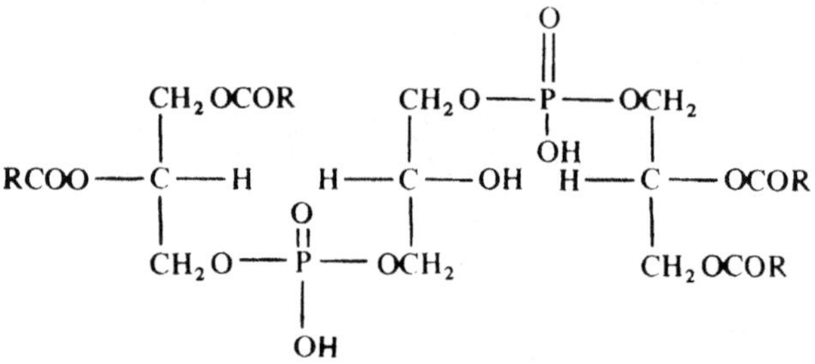

1′,3′ - di - *O* - (3 - *sn* - phosphatidyl) - *sn* - glycerol

(26)

By far the most common of the phosphoglycerides are the **phosphatidyl esters** in which phosphatidic acid (**24**) is esterified with another hydroxy compound to produce a diester of phosphoric acid (a monohydrogen phosphate). The phosphatidyl esters are still acidic by reason of the remaining ionisable hydrogen atom. The hydroxy compounds most commonly incorporated include choline ($HOCH_2CH_2\overset{+}{N}Me_3$) to give **phosphatidylcholines** (PC, **27**), ethanolamine ($HOCH_2CH_2NH_2$) to give **phosphatidylethanolamines** (PE, **28**), serine ($HOCH_2CH(NH_2)COOH$) furnishing **phosphatidylserines** (PS, **29**), and inositol (hexahydroxycyclohexane) giving **phosphatidylinositides** (PI, **30**). Lecithin is an old name for the phosphatidylcholines and the term cephalin was used to describe mixtures of phosphatidylethanolamines and phosphatidylserines. Phosphatidylinositol also exists in forms in which one or more of the free hydroxy groups on the inositol is/are linked to additional phosphoric acid groups or to sugar units. All the natural

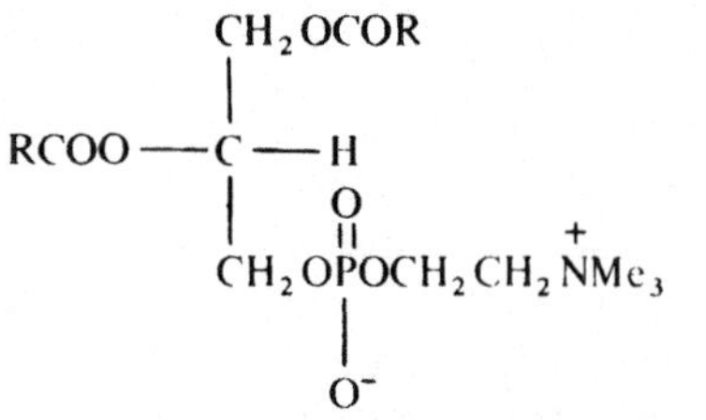

3 - *sn* - phosphatidylcholine PC (27)

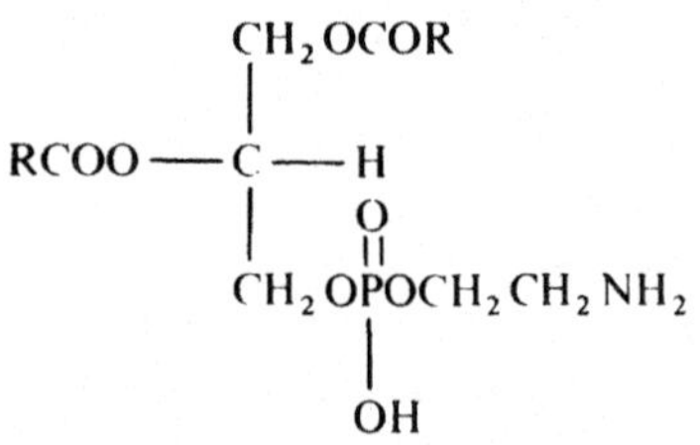

3 - *sn* - phosphatidylethanolamine PE (28)

phosphatidyl esters are based on *sn*-glycerol 3-phosphate, i.e. on one enantiomeric form of this compound.

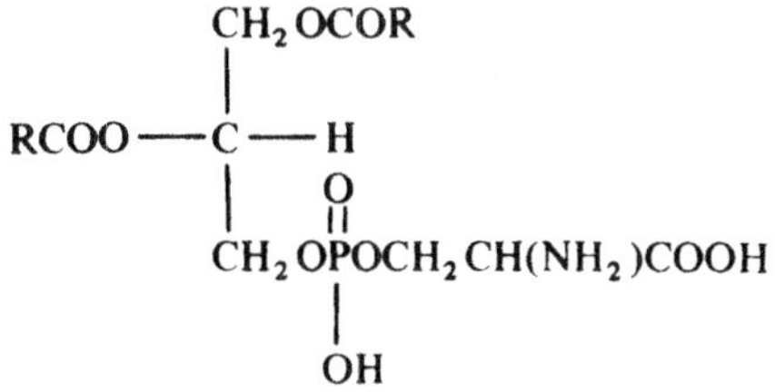

3 - *sn* - phosphatidylserine PS **(29)**

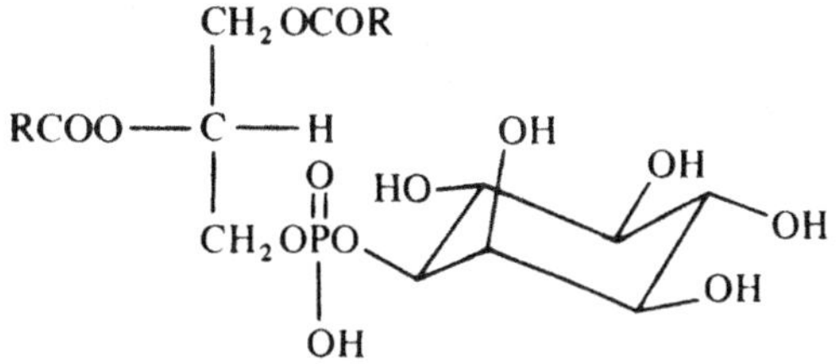

3 - *sn* - phosphatidyl - 1′ - myo - inositol PI **(30)**

(also occurs as the 4′ - phosphate and the 4′, 5′ - diphosphate)

Complete hydrolysis of a phosphatidyl ester is easily effected in acidic media. With alkali the acyl groups are split off readily leaving intermediates such as glycerophosphorylcholine (etc.) which are hydrolysed only slowly to choline (etc.) and the isomeric mixture of α- and β-glycerophosphoric acids.

$$\text{PC} \xrightarrow[\text{hydrolysis}]{\text{mild}} \begin{cases} \text{fatty acids} \\ + \text{GPC} \xrightarrow[\text{hydrolysis}]{\text{strong}} \text{choline} + \alpha\text{- and } \beta\text{-GPA} \end{cases}$$

Enzymes operate more specifically and effect selective hydrolysis of one or other of the four ester bonds. Each of the enzymes phospholipase-A^1, phospholipase-A^2, phospholipase-C, and phospholipase-D cleave a phosphatidyl ester into two products as indicated in (**31**). (The material first designated phospholipase-B was found to be a mixture of phospholipase-A^1 and phospholipase-A^2.) Phosphatidyl esters which have been deacylated once and retain one acyl group are called **lysophosphatidyl esters**.

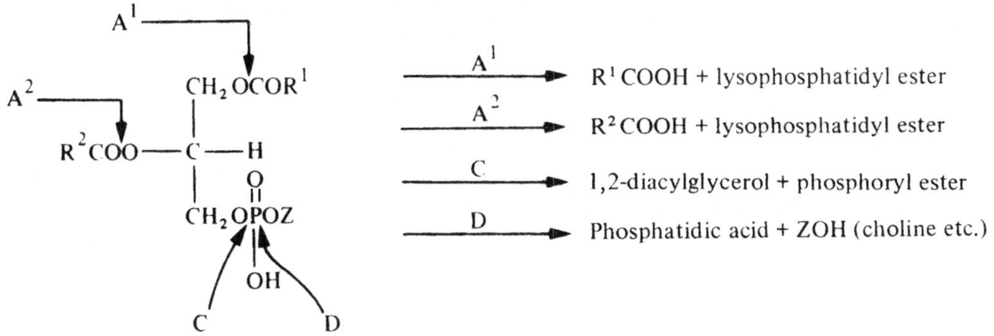

(31) (Z = the remainder of the appropriate hydroxy compound)

6. Ether lipids

Ether lipids are variants of the structures already discussed. In these the acyl group attached to C-1 is replaced by an alkyl group (which may be saturated or unsaturated) or by an alkenyl group having *cis*(*Z*) unsaturated between C-1 and C-2 (in addition to other possible unsaturated centres). There are four types of such compounds (**32–35**) depending on whether the ether link is introduced into a triacylglycerol or a phosphatidyl ester. The alkyl and alken-1-yl groups have the same chain-lengths as the common fatty acids.

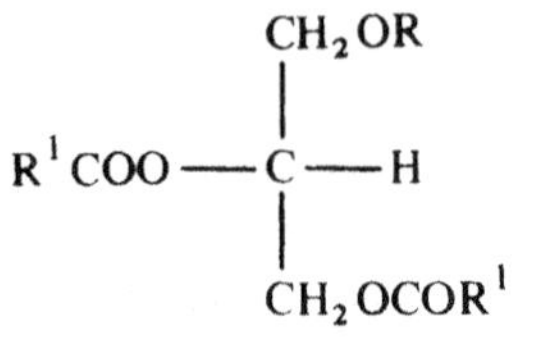

(32)

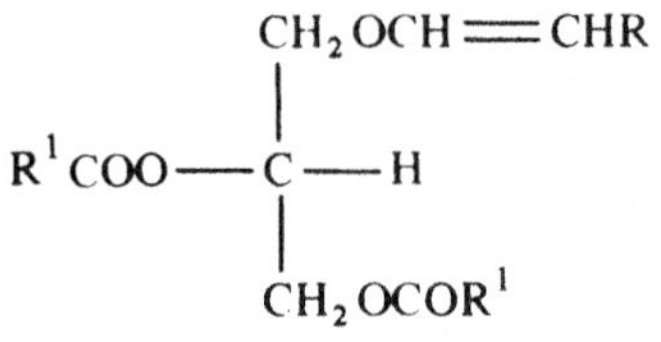

(33)

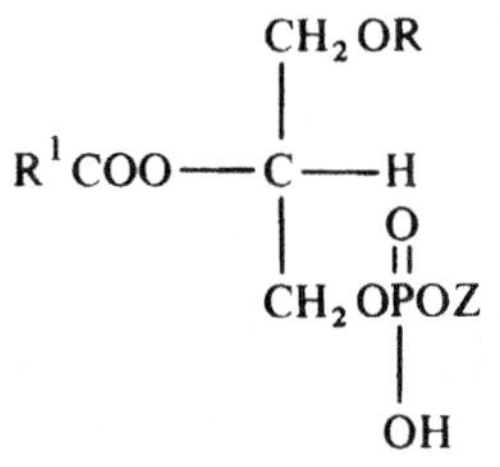

(34)

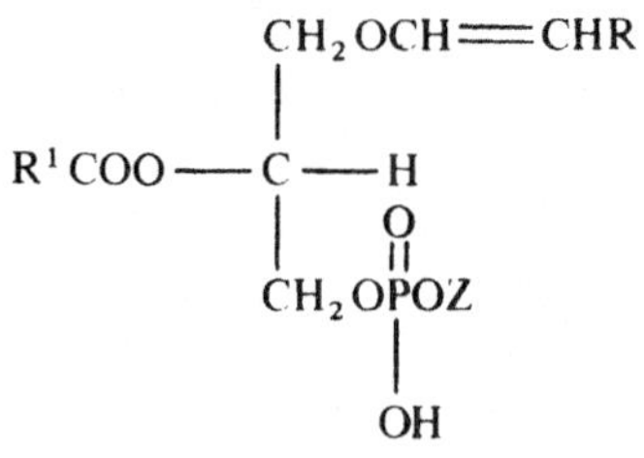

(35)

(Z represents the residue from choline, ethanolamine, etc.)

The **alkyldiacylglycerols** (**32**) are common constituents of some marine oils. When hydrolysed the acyl groups are split off and glycerol ethers such as hexadecylglycerol, octadecylglycerol, and octadec-9-enylglycerol remain. The vinyl ethers (**33** and **35**), whilst resistant to hydrolysis by alkali or reduction by lithium aluminium hydride, are hydrolysed by dilute acid to furnish long-chain aldehydes. They were originally called plasmalogens because they were found in blood and gave aldehydes on acidic hydrolysis.

$$R^1OCH{=}CHR^2 \longrightarrow R^1OH + [R^2CH{=}CHOH] \longrightarrow R^2CH_2CHO$$

7. Sphingolipids (ceramides, cerebrosides, gangliosides, phosphosphingolipids, glycosphingolipids)

In contrast to all the *ester* lipid types already discussed, the sphingolipids contain fatty acids combined as *amides* of long-chain compounds containing an amino group and two or more hydroxyl groups.

These natural bases are C_{12}–C_{22} compounds of two types. Those (**36**) containing three functional groups and at least one double bond (of *t* configuration) are mainly of animal origin and were known as **sphingosines**. The **phytosphingosines** (37) with four functional groups are mainly of plant origin. All the chiral centres in both kinds of molecule have D-configuration.

$$CH_3(CH_2)_nCH\overset{t}{=}CHCH(OH)CH(NHX)CH_2OH$$

X = H sphingosine (**36**)

X = COR ceramide (**38**)

$$CH_3(CH_2)_mCH(OH)CH(OH)CH(NHX)CH_2OH$$

X = H phytosphingosine (37)

X = COR ceramide (39)

New nomenclature relates these compounds to the C_{18} dihydro derivative of **36** which is designated sphinganine. Compounds **36** (C_{18}) and **37** (C_{20}) are then sphing-4*t*-enine and 4-hydroxyicosasphinganine respectively.

In their N-acylated forms these long-chain amines are called **ceramides** (**38** and **39**). They occur naturally as structures in which the primary hydroxyl group [but not the secondary hydroxyl group(s)] is associated either with a sugar unit or with phosphate esters such as those already encountered in the phosphatidyl esters.

In some sphingolipids the ceramide is linked through its primary hydroxyl group with a sugar moiety which may be simple (galactose or glucose) or complex. Monoglycosylceramides are known as **cerebrosides** and the more complex compounds as gangliosides. Though first isolated from brain—as their name implies—cerebrosides occur widely throughout the animal and vegetable kingdoms. The associated fatty acids are mainly alkanoic and 2-D-hydroxyalkanoic acids. The more complex gangliosides, present in brain lipids, erythrocite membranes, and other tissues, are found in high concentration in the ganglion cells of the central nervous system.

Phosphosphingolipids of animal origin also contain phosphorylcholine or phosphorylethanolamine and are called **sphingomyelins** (**40**). Structure (**41**) represents a phosphosphingolipid of plant origin.

$$CH_3(CH_2)_{12}CH\overset{t}{=}CHCH(OH)CH(NHCOR)CH_2O\underset{\displaystyle O^-}{\underset{|}{\overset{\displaystyle O}{\overset{\|}{P}}}}OCH_2CH_2\overset{+}{N}Me_3 \qquad (\mathbf{40})$$

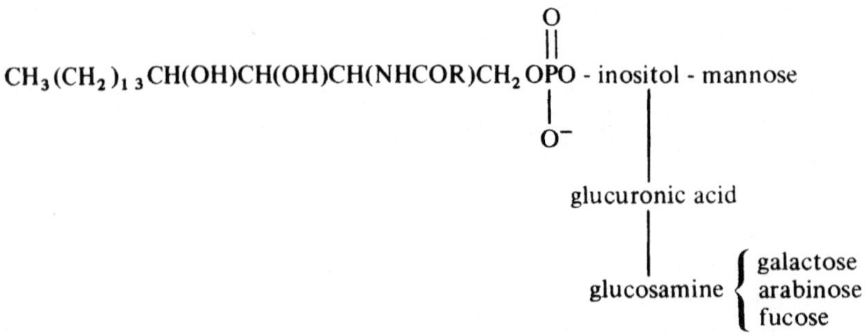

(41)

Though chemically distinct, the phosphatidyl esters and the sphingolipids have some similar physical properties resulting from the association of two long acyl chains and a polar head group in each molecule.

GENERAL REFERENCES

GUNSTONE, F. D., *Fats and Oils: Chemistry and Technology* (ed. R. J. Hamilton and A. Bhati), Applied Science, London, 1980, p. 47.

GURR, M. I. and A. T. JAMES, *Lipid Biochemistry: An Introduction*, 3rd ed., Chapman & Hall, London, 1980, p. 18.

HITCHCOCK, C., *Recent Advances in the Chemistry and Biochemistry of Plant Lipids* (ed. T. Galliard and E. I. Mercer), Academic Press, London, 1975, p. 1.

PRYDE, E. H., *Fatty Acids* (ed. E. H. Pryde), American Oil Chemists' Society, Champaign, 1979, p. 1.

SMITH, C. R., *Fatty Acids* (ed. E. H. Pryde), American Oil Chemists' Society, Champaign, 1979, p. 29.

CHAPTER 2

The Separation and Isolation of Fatty Acids and Lipids

A. INTRODUCTION

Natural lipids are usually mixtures of several classes of lipids and each class contains a wide range of fatty acids. Separation procedures are therefore important for the isolation of pure acids which are usually more easily obtained from natural sources than by chemical synthesis. Also important for analytical purposes are the separation of individual lipid classes, of individual lipids within a class, and of the fatty acids from a single source. In this chapter the emphasis will be on preparative separation leading to the isolation of pure compounds or of concentrates. Analytical aspects are discussed in Chapter 3, though these two aspects of separation procedure are not easily disentangled.

B. DISTILLATION

Since the effectiveness of distillation depends on differences of boiling point, chain-length is more important than the degree of unsaturation. It is possible to separate the esters of C_{12}, C_{14}, C_{16}, C_{18}, and C_{20} acids from each other, and distillation is commonly used for this purpose. On the other hand, stearate (18:0), oleate (18:1), linoleate (18:2), and linolenate (18:3) are not separable by distillation.

The most widely used distillation procedure is fractional distillation of methyl esters under reduced pressure (0.1–1.0 mm). Even under these conditions, moderately high temperatures are required (bath temperatures around 200°) and the more highly unsaturated acids, especially those with conjugated unsaturation or those longer than C_{18}, are prone to polymerisation, cyclisation, and stereomutation of double bonds. Heated columns packed with glass helices or some form of metal packing are in common use despite the disadvantage of a significant hold-up and pressure-drop through the column. Spinning bond columns do not suffer from these disadvantages.

Distillation at still lower pressures has been used in the isolation of some highly unsaturated acids and is particularly valuable in polymerisation studies to separate monomeric, dimeric, and polymeric material.

C. CRYSTALLISATION[1]*

Fatty acids can be concentrated from natural mixtures by crystallisation from appropriate organic solvents. Repeated crystallisation may give an improved product, but the isolation of pure compounds (>99%) usually involves the application of several different separation procedures (Section I).

Saturated acids are usually solid at room temperature (Table 1.2) and can be crystallised at room temperature or at cold-room temperature. Crystals are separated from mother liquor by any conventional filtration procedure.

Unsaturated acids with lower melting points and higher solubilities must be crystallised at temperatures (between 0 and −90°) at which the unsaturated acids are solid. Filtration must also be carried out at an appropriate low temperature.

Crystallisation is a mild procedure suitable for the polyene acids which are so easily oxidised or otherwise modified at elevated temperatures. Separation of saturated and monoene acids from one another by crystallisation is good: separation of polyene acids from one another is less satisfactory. Useful solvents include methanol, ether, petroleum ether, and acetone usually at a dilution of 5–10 ml of solvent per gram of acid.

D. UREA FRACTIONATION[2]

Urea normally crystallises in tetragonal form, but in the presence of certain aliphatic compounds it forms hexagonal prisms containing some of the aliphatic material. These prisms are built up from urea: six molecules form a unit cell 11.1×10^{-10} m long and 8.2×10^{-10} m in diameter containing a channel in which an open-chain molecule may be held so long as it fulfills certain dimensional qualifications. It must not be too short or it will not be held within the channel, and it must not be too wide if it is to fit into the free space, variously estimated at between 4.0 and 6.0×10^{-10} m. Many straight-chain acids and their esters satisfy these conditions and thus readily form complexes (adducts, inclusion compounds) with urea.

Saturated acids form stable complexes more readily than do the unsaturated acids, and oleic acid enters into these inclusion compounds more readily than do the polyunsaturated acids. In practice urea and mixed acids are dissolved in hot methanol or urea and methyl esters in a hot methanol–ethanol mixture. The solution is crystallised at room temperature or at 0°. The adduct and mother liquor will furnish the acids or esters when mixed with water and extracted with ether or petroleum ether in the usual way.

This procedure is employed for two purposes. It separates straight-chain acids or esters from branched-chain or cyclic compounds with the former concentrating in the adduct and the latter in the mother liquor. Urea fractionation is also used to separate acids or esters of differing unsaturation and is an important process in the isolation of pure methyl oleate, linoleate, or linolenate (Section I).

E. ADSORPTION CHROMATOGRAPHY ON SILICA[3,4]

Compounds differing in polarity can be separated by adsorption chromatography on silica carried out in columns or in the thin-layer mode. Thin-layer separations may be analytical or

* Superscript numbers refer to References at end of chapter.

preparative (up to 100 mg per 20 × 20 cm plate depending on the ease of separation). Separated components can be made visible with a destructive spray (suitable for analytical purposes) or with a non-destructive spray (necessary for preparative purposes). It is less easy to follow a column separation and fractions may have to be collected arbitrarily and then examined by a suitable analytical procedure.

Compounds of very different polarity, such as esters of ordinary long-chain acids and esters of hydroxy acids, are easily separated in this way, but the small difference in polarity between esters of differing unsaturation can also be exploited. Methyl stearate, oleate, linoleate, and linolenate are eluted from silica in that order and this can be exploited to increase the purity of an already fairly pure ester. It is not suitable for the initial enrichment of individual esters. Petroleum ether (b.p. 40–60°) containing increasing proportions of diethyl ether is frequently used as eluting solvent.

Lipids of different classes are separable by chromatography on silica, but this is more commonly used as an analytical procedure (Chapter 3).

F. ADSORPTION CHROMATOGRAPHY ON SILICA IMPREGNATED WITH SILVER NITRATE

Silver nitrate interacts with the π-electrons of olefinic compounds, thereby modifying their chromatographic behaviour. A mixture of silica and silver nitrate (10–30 %) is used as a column or a thin layer. *cis*-Olefinic esters are retarded more than their *trans*-isomers because of their stronger interaction with silver ions and all *cis*-unsaturated esters are held back in proportion to the number of double bonds they contain. Lipids of a single class (e.g. triacylglycerols) can be similarly separated on the basis of their degree of unsaturation. Ion exchange columns charged with silver ions have been used in a similar way.

G. HIGH-PERFORMANCE LIQUID CHROMATOGRAPHY

HPLC is the newest of the chromatographic procedures to be used for the isolation and purification of fatty acids and lipids and already shows great promise. It can be used in both an analytical and a preparative mode and is more efficient and quicker than conventional column chromatography.

Acids or esters differing markedly in polarity are easily separated on silica columns and such a procedure can be used to prepare methyl ricinoleate (hydroxyoleate) from the methyl esters of castor oil.

Esters showing only small differences in polarity such as oleate, linoleate, and linolenate are better separated on reversed phase columns. Up to 10 g quantities of pure ester (>99 %) can be obtained from suitable natural mixtures in 20–30 min. Though the scale is modest, the procedure is very efficient.

H. GAS–LIQUID CHROMATOGRAPHY[5–7]

Gas–liquid chromatography is most often employed for analytical purposes (Chapter 3), but it can be used preparatively for small amounts of material (μg to mg).

I. ISOLATION OF SOME OF THE MORE COMMON UNSATURATED ACIDS[8,9]

Oleic acid or ester is best obtained from a vegetable source in which oleate is virtually the only monoene acid present. Olive oil is a useful starting material. Animal fats are less suitable since they are more likely to contain isomeric octadecenoic acids. The following procedure has been recommended: (a) urea fractionation to reduce saturated acids to below 1 %—two treatments may be necessary, (b) crystallisation from petroleum ether at −40° three or four times until oleic acid exceeds 99 %, and (c) chromatography of acids or methyl esters to remove coloured impurities and oxidised products.

Linoleic acid or ester can be obtained from any linoleic-rich oil which has less than 1 % of linolenic acid by the following steps: (a) urea fractionation to reduce the combined content of saturated and oleic acids in the mother liquor to <5 %, (b) repeated low temperature crystallisation from acetone at −75° until the precipitate contains >90 % linoleic acid, (c) further urea fractionation to reduce the combined content of saturated and oleic acid to below 0.8 %, and (d) low temperature crystallisation at −75° to give pure acid (>99 %) which is finally eluted from a column of silica.

α-Linolenic acid or ester is usually prepared from linseed oil, but this is a tedious purification because of the difficulty of separating linoleic from linolenic acid. The proportion of linolenic acid (55–60 %) in linseed oil is raised to 65–75 % by low-temperature crystallisation from acetone at −75° and then to 80–90 % by urea fractionation. This material is further upgraded by careful elution from a column of silica.

Arachidonic acid is obtained from pig liver lipid by distillation of the methyl esters after chromatographic removal of cholesterol, followed by urea fractionation and silver ion chromatography.

REFERENCES

1. J. B. Brown and D. K. Kolb, *Progress in the Chemistry of Fats and Other Lipids*, 1955, **3**, 57.
2. H. Schlenk, *Progress in the Chemistry of Fats and Other Lipids*, 1954, **2**, 243.
3. D. C. Malins, *Progress in the Chemistry of Fats and Other Lipids*, 1966, **8**, 301.
4. R. A. Stein, *Progress in the Chemistry of Fats and Other Lipids*, 1966, **8**, 373.
5. R. G. Ackman, *Progress in the Chemistry of Fats and Other Lipids*, 1972, **12**, 165.
6. A. Kuksis, *Handbook of Lipid Research* 1. *Fatty Acids and Glycerides* (ed. A. Kuksis), Plenum Press, New York, 1978, p. 1.
7. J. J. Myher, *Handbook of Lipid Research* 1. *Fatty Acids and Glycerides* (ed. A. Kuksis), Plenum Press, New York, 1978, p. 123.
8. O. S. Privett, *Progress in the Chemistry of Fats and Other Lipids*, 1968, **9**, 407.
9. F. D. Gunstone, J. McLaughlin, C. M. Scrimgeour and A. P. Watson, *J. Sci. Fd. Agric.* 1976, **27**, 675.

GENERAL REFERENCE

Sonntag, N. O. V., *Fatty Acids* (ed. E. H. Pryde), American Oil Chemists' Society, Champaign, 1979, p. 125.

CHAPTER 3

The Analysis of Fatty Acids and Lipids

A. THE NATURE OF THE PROBLEM

Before the development of modern methods of chromatography it was only possible to measure the bulk properties of a lipid sample such as its mean unsaturation (iodine value), the average molecular weight of its component acids (saponification equivalent), or the percentage of an element such as nitrogen or phosphorus or sulphur. For further information the lipid mixture was separated into fractions before carrying out the above measurements. Lipid analysis can now be conducted more satisfactorily through the use of chromatography and of enzymic deacylation procedures.

The modern study of a lipid sample can be conducted at several analytical levels. It is possible to determine, both qualitatively and quantitatively, the nature of the acids, alcohols, or aldehydes associated with the total sample under investigation or to separate the lipid extract into its several types before examining the component acids, etc. Additionally, the fatty acids attached to each of the three carbon atoms in a triacylglycerol or the two carbon atoms in a phosphoglyceride can be distinguished by enzymatic processes. Finally, by a combination of chromatography and enzymic deacylation, individual molecular species can sometimes be quantified.

The complexity of the problem derives from the large number of molecular species which may be present in a natural mixture. For example, phosphoglycerides contain two acyl groups in each molecule and if the lipid contains ten acids then there could be 100 different molecular types. The situation is more complex with the triacylglycerols. The number of possible compounds from n fatty acids is n^3 if all isomers (including enantiomers) are distinguished, $(n^3 + n^2) \div 2$ if optical isomers are not distinguished, and $(n^3 + 3n^2 + 2n) \div 6$ if no isomers are distinguished. These numbers soon become large so that if there are five (125, 75, 35), ten (1000, 550 and 220), or twenty component acids (8000, 4200, 1540) they have the values indicated in parenthesis. Most seed oils contain 5–10 different acids, but animal fats are more complex with 10–40 acids. The number of acids found in a natural lipid sample depends to some extent on the thoroughness with which the sample has been examined. For example, more than 140 acids have been identified in the much-studied cow milk fat with most of these present only at very low concentrations.

The results obtained from studies of this kind show interesting patterns of association and distribution. They reveal, for example, that fatty acids are not distributed at random, but that there is a marked selectivity in acyl chains associated with each hydroxyl group.

B. COMPONENT ACIDS, ALCOHOLS, AND ALDEHYDES[1,2]*

Mixtures of long-chain acids, alcohols, or aldehydes derived from natural lipids differ mainly in chain-length and extent of unsaturation though they may, less commonly, contain branched-chains, cyclic systems, or additional functional groups. After appropriate derivatisation, these mixtures are analysed quantitatively by gas chromatography. With more complex mixtures, or if a more refined analysis is required, gas chromatography is carried out on more than one phase or is combined with other separation procedures such as silver ion chromatography (Chapter 2). Capillary columns give improved resolution and may show an increased number of components.

For gas chromatography, carboxylic acids are usually converted to methyl esters; alcohols to acetates, trifluoroacetates, or trimethylsilyl ethers; aldehydes are examined as such, as dimethylacetals, or after conversion to alcohols or acids; and sphingosine bases are converted to trimethylsilyl ethers or to aldehydes by oxidation with sodium metaperiodate thus:

$$RCH{=}CHCH(OH)CH(NH_2)CH_2OH \longrightarrow RCH{=}CHCHO$$

The identification of components appearing as peaks in a chromatogram is normally achieved by comparison with the known chromatographic behaviour of such compounds. This is sufficient for the major and more common esters, but may be misleading for minor and less common constituents. In the extreme case it may be necessary to isolate and identify particular components. Both these processes are achieved in systems combining the separating power of gas chromatography with the identifying power of mass spectrometry. Some of the chemical and physical methods of identification will be discussed in other connections, but they are not drawn together in this book since this is not a common problem for the food technologist.[3,4] Reference to the use of HPLC is given at the end of the following section.

C. COMPONENT LIPIDS[5,6]

The recognition of lipid types in a mixture and their quantitative determination is achieved mainly by chromatographic separation. The most widely used procedures are based on adsorption chromatography with silica or acid-washed florisil. Column or thin-layer techniques are employed and quantitative analysis is achieved by weighing or by densitometry, fluorimetry (thin-layer systems only), or gas chromatography of component acids after addition of an internal standard. Neutral lipids are eluted from a column with chloroform, glycolipids with acetone, and phospholipids with methanol.

The neutral lipids are further separated on a second silica column: hydrocarbons are eluted with hexane alone, cholesterol esters with 2% ether in hexane, triacylglycerols with 5% ether, cholesterol and diacylglycerols with 15% ether, and monoacylglycerols with ether alone. Similar separations are achieved on thin layers of silica using mixtures of hexane (90–70%) and ether (10–30%) usually with 1–2% of added formic or acetic acid.

Complex lipids are further separated on columns with chloroform containing increasing proportions of methanol or on thin layers with appropriate solvent mixtures. With chloroform–methanol mixtures acidic lipids are eluted in the order: cardiolipin, phosphatidic acid, cerebroside sulphate, phosphatidylglycerol, phosphatidylserine, phosphatidylinositol, and di- and

* Superscript numbers refer to References at end of chapter.

triphosphoinositides. Non-acidic lipids are eluted with the same solvent systems in the order: ceramide, ceramide monohexoside and glycosyldiglycerides, phosphatidylethanolamine, phosphatidylcholine, lysophosphatidylethanolamine and ceramide dihexoside, sphingomyelin, and lysophosphatidylcholine.

Developments in gas chromatography make it increasingly possible to separate the more volatile lipid classes from one another after suitable derivatisation (Section E).

Analytical high-performance liquid chromatography (HPLC) is useful for the separation of fatty acids—suitably derivatised—and of various lipid classes.[7] It may be used in the adsorption mode using silica columns and eluting with solvents of low to medium polarity such as those lying between petroleum ether and ethyl acetate. Silver ions may be incorporated into the column to improve the separation of unsaturated components. In the more commonly employed reversed-phase system a hydrophobic stationary phase competes with hydrophilic mobile phases such as methanol and acetonitrile for the lipid components. The hydrophobic C_{18} packing material consists of porous silica, the surface of which is coated with a monolayer of saturated hydrocarbon. The coating is accomplished by covalently bonding C_{18} alkylsilane to silanol sites on the silica surface. The application of HPLC to fatty acid and lipid analyses is restricted by the limitations of the ultraviolet and refractive index detectors for these kinds of compounds.

D. ENZYMATIC DEACYLATION OF LIPIDS[8,9]

1. Hydrolysis with pancreatic lipase

Pig pancreatic lipase promotes the hydrolysis of acyl groups attached to the two primary hydroxyl functions only, so that the final products of lipolysis are 2-monoacylglycerols and fatty acids liberated from the C-1 and C-3 positions. Esters of acids with chain-branching or unsaturation close to the carboxyl group (e.g. 16:1 3*t* and C_{20} and C_{22} polyenes with unsaturation starting at Δ4 or Δ5) are hydrolysed more slowly than esters of the common acids: esters of short-chain acids are hydrolysed more quickly. The method is thus reliable for almost all natural triacylglycerols except fish oils (rich in C_{20} and C_{22} polyenes) and milk fats (rich in short-chain acids).

triacylglycerols $\rightleftharpoons$ 1,2- and 2,3-diacylglycerols and fatty acids $\rightleftharpoons$ 2-monoacylglycerols and fatty acids

Hydrolysis to the 50–60% level is effected in a buffered solution (pH 8.5) at 37° in the presence of bile salts and calcium ions. 2-Monoacylglycerols are isolated chromatographically and their component acids determined by gas chromatography—usually as methyl esters after transesterification. Fatty acids attached at C-1/2/3 together are known from the composition of the whole oil, fatty acids attached to C-2 from the 2-monoacylglycerols, and the C-1/3 fatty acids by difference. It is not possible by this procedure, however, to distinguish the C-1 and C-3 acids.

This was the first enzymatic procedure to be applied to lipid analysis. It showed that acyl groups attached to C-2 were generally different from those attached to C-1/3. In vegetable fats, C-2 is acylated almost entirely with unsaturated C_{18} acids (oleic, linoleic, linolenic) whilst saturated and long-chain unsaturated acids (20:1, 22:1) appear at C-1/3 and rarely at C-2. Some examples are given in Table 3.1. The distribution of acids in animal fats and fish oils is less consistent. The pig and wild boar are unusual in that palmitic acid predominates at C-2. In other land animals oleic and hexadecenoic acid are enriched at this position, whilst stearic acid consistently concentrates at C-1/3. In fish oils, the polyene (20:5, 22:6) and lower saturated acids (14:0, 16:0) concentrate at C-2 and the monoene acids, particularly 20:1 and 22:1, at C-1/3.

TABLE 3.1. *The Component Acids of Some Vegetable and Animal Fats (TG) and of the 2-Monoacylglycerols (MG) derived from them by Lipolysis*[a]

		16:0	16:1	18:0	18:1	18:2
Almond	TG	7		2	70	21
	MG	1		0	64	34
Poppy	TG	10		2	11	76
	MG	1		0	9	89
Palm	TG	44		6	39	9
	MG	11		2	65	22
Rape[b]	TG	3		2	22	15
	MG	1		0	37	37
Pig	TG	28	3	15	42	9
	MG	72	4	4	12	3
Sheep	TG	27	3	27	35	2
	MG	14	5	9	58	6

[a]Taken from F. D. Gunstone, *Comprehensive Organic Chemistry* (ed. Sir Derek Barton and W. D. Ollis), Pergamon, 1979, p. 643.
[b]Also 18:3 (14, 23), 20:1 (15, 0), and 22:1 (28, 1). The two figures in parenthesis indicate the percentage of acid in triacylglycerol and 2-monoacylglycerol, respectively.

Further conclusions from lipolysis results can only be obtained on the basis of certain assumptions. It is not uncommon to calculate glyceride composition *assuming* (i) that acyl distribution at C-1 and C-3 is the same and that (ii) that acyl groups present at each of the three positions are randomly associated with each other. These assumptions may be true of some natural fats and oils, but they do not apply to segregated fractions derived from natural mixtures by crystallisation or chromatography.

2. Regio- and stereospecific analysis of triacylglycerols

Methods of analysing the acyl chains at each of the three glycerol positions were pioneered by Brockerhoff and the most popular of these is described.

Triacylglycerol (**A**) is treated for 1 min with ethylmagnesium bromide which reacts in a random manner with the ester linkages. $\alpha\beta$-Diacylglycerols (**B**), separated from $\alpha\alpha'$-diacylglycerols and tertiary alcohols by thin-layer chromatography, are treated with phenyldichlorophosphate to give phosphatidylphenols of two types (**C** and **C′**). Only the *sn*-3-phosphatidylphenol is hydrolysed by phospholipase A so that the reaction product comprises unreacted phosphatidylphenol (C′) containing the acyl groups attached to C-2 and C-3, free acid from C-2 (**D**), and lysophosphatidylphenol (**E**) containing the C-1 acyl groups. The C-2 acyl groups are also determined by pancreatic lipase hydrolysis. Although the procedure requires two synthetic steps, two enzymic hydrolyses, two thin-layer separations, five transesterifications, and five gas chromatographic analyses, consistent and reliable results can be obtained. (Some advantage is now claimed for using the phosphatidylcholines rather than phosphatidylphenols.) The C-1 fatty acids are found from the analysis of **E**, the C-2 acids from the 2-monoacylglycerols (**F**), and those at C-3 are calculated from 3**A**-**E**-**F** or **2C**-**F**.

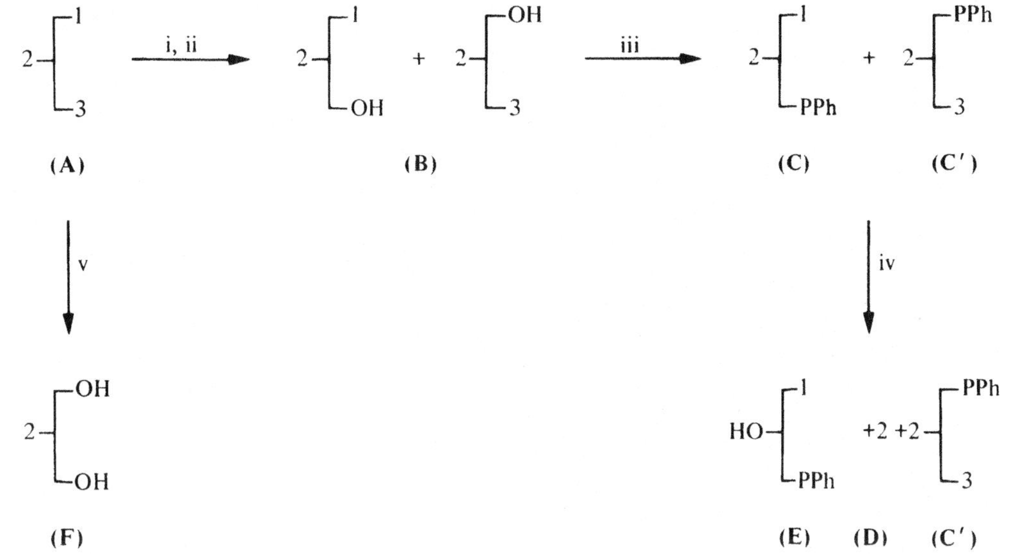

Brockerhoff's stereospecific analysis of triacylglycerols. The symbols indicate the glycerol backbone and the acyl groups attached to the *sn*-1, -2 and -3 positions; PPh = $-OPO(O^-)OPh$
i, EtMgBr; ii, thin-layer chromatography; iii, phenyldichlorophosphate, pyridine; iv, phospholipase A; v, pancreatic lipase.

SCHEME 3.1

The results obtained by this and similar procedures have shown that in most cases—though not in all—natural triacylglycerols have different fatty acids associated with each hydroxyl group. This information does not indicate the individual triacylglycerols present and these can be calculated from the results only on the *assumption* that each set of acyl groups is distributed at random. Some typical results are given in Table 3.2.

3. Stereospecific analysis of phosphoglycerides

Phospholipase A_2, from snake venom, selectively deacylates practically all phosphoglycerides except the phosphatidylinositols. The natural compounds based on *sn*-phosphatidic acids furnish a lysophosphoglyceride with release of the fatty acids from C-2. Analysis of the liberated acids and of those remaining in the lysophosphoglyceride provide information about the acids at C-2 and C-1 respectively. Results show fairly consistently that saturated acids predominate at C-1 and

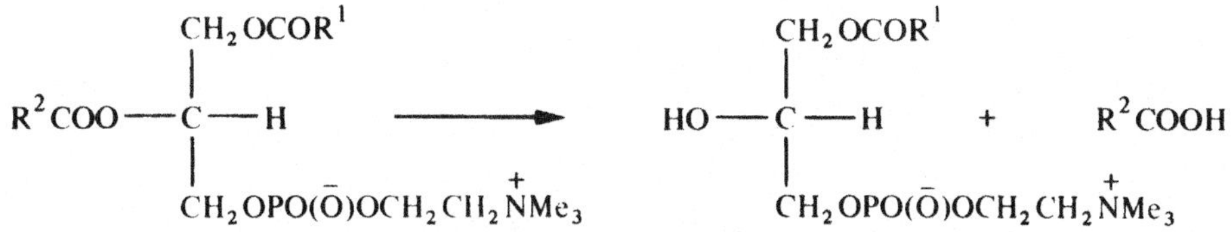

TABLE 3.2. *Stereospecific Analysis of Some Natural Fats*[a]

		16:0	18:0	18:1	18:2	Other acids[b]
Vegetable fats						
Soybean	1	14	6	23	48	18:3 (9, 7, 8)
	2	1	tr	22	70	
	3	13	6	28	45	
Cocoa butter	1	34	50	12	1	20:0 (1, 0, 2)
	2	2	2	87	9	
	3	37	53	9	tr	
Rape	1	4	2	23	11	18:3 (6, 20, 3); 20:1 (16, 2, 17); 22:1 (35, 4, 51)
	2	1	0	37	36	
	3	4	3	17	4	
Animal fats						
Man	1	39	10	33	3	14:0 (4, 11, 1); 16:1 (5, 11, 4)
	2	10	2	50	9	
	3	25	9	51	5	
Pig	1	16	21	44	12	14:0 (2, 4, tr); 16:1 (3, 4, 3)
	2	59	3	17	8	
	3	2	10	65	24	
Ox	1	41	17	20	4	14:0 (4, 9, 1); 16:1 (6, 6, 6)
	2	17	9	41	5	
	3	22	24	37	5	
Cow (milk)	1	36	15	21	1	4:0 (5, 3, 43); 6:0 (3, 5, 11) 8:0 (1, 2, 2); 10:0 (3, 6, 4) 12:0 (3, 6, 4); 14:0 (11, 20, 7); 16:1 (3, 2, 1)
	2	33	6	14	3	
	3	10	4	15	0	
Chicken	1	25	6	33	14	16:1 (12, 7, 12)
	2	15	4	43	23	
	3	24	6	35	14	
Herring gull	1	22	13	41	7	16:1 (4, 3, 5); 20:1 (7, 6, 7); 22:1 (3, 4, 5)
	2	15	9	48	11	
	3	17	7	46	9	
Cod	1	15	6	28	2	14:0 (6, 8, 4); 16:1 (14, 12, 14); 20:1 (12, 7, 17); 22:1 (6, 5, 7); 20:5 (2, 12, 13); 22:5 (1, 3, 1); 22:6 (1, 20, 6).
	2	16	1	9	2	
	3	7	1	23	2	

[a]Taken from F. D. Gunstone, *Comprehensive Organic Chemistry* (ed. Sir Derek Barton and W. D. Ollis), Pergamon, 1979, p. 645.
[b]The three figures in parenthesis indicate the percentage of acid at the 1, 2, and 3 positions, respectively.

unsaturated acids at C-2, though phosphoglycerides with two saturated or two unsaturated acyl groups are known. It is only possible to calculate the component phosphoglycerides from these results if it is *assumed* that the two sets of acyl groups are distributed statistically. Some typical results are given in Table 3.3.

E. ATTEMPTS TO SEPARATE INDIVIDUAL LIPID SPECIES

In this section reference is made to attempts to isolate individual lipids or to obtain mixtures simple enough to be analysed unambiguously. Suitable procedures have been developed for triacylglycerols and these can be applied to ester waxes, mono- and diacylglycerols, and to phospholipids, in some cases after modification of the lipid.

TABLE 3.3. *The Distribution of Major Component Acids in Some Phosphatidylcholines*

Fatty acid position	16:0 1	16:0 2	18:0 1	18:0 2	18:1 1	18:1 2	18:2 1	18:2 2	20:4 1	20:4 2	Other acids[b]
Salmon	37	10	3	0	8	23	—	—	—	—	20:5 (14, 17) 22:6 (33, 46)
Egg	61	5	25	2	10	59	2	26	0	6	
Bovine liver	19	1	46	0	19	17	3	16	2	18	22:1 (2, 15) 22:4 (1, 7) 22:5 (2, 10) 22:6 (0, 6)
Bovine milk	47	35	20	3	23	32	3	11	—	—	14:0 (6, 13)
Human serum	66	3	26	0	6	22	1	47	0	12	20:3 (0, 5) 22:6 (0, 6)

[a]Taken from F. D. Gunstone, *Comprehensive Organic Chemistry* (ed. Sir Derek Barton and W. D. Ollis), Pergamon, 1979, p. 645.
[b]The two figures in parenthesis indicate the percentage of acid at the 1 and 2 positions, respectively.

Triacylglycerols containing unusual acids, such as milk fats with short-chain acids or seed oils with oxygenated acids, can sometimes be separated on thin layers of silica, but silver ion chromatography is more generally useful and has been successfully applied to unsaturated vegetable oils with glycerides having 0-9 double bonds and to fish oils with each acyl group having

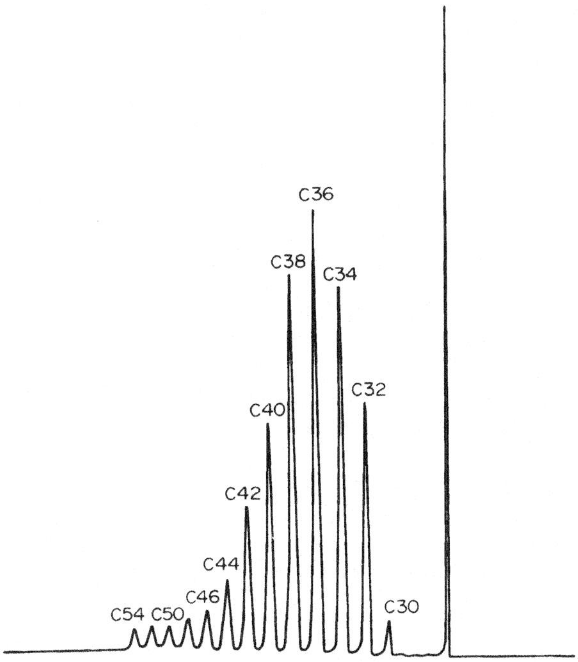

FIG. 3.1. GLC of coconut oil. Column: 45 cm × 2 mm internal diameter packed with Dexsil (1 %) on 100/120 Supelcoport. Injection heater at 350°C ± 1°. Nitrogen flow rate 60 ml per min. Column temperature 230–360° at 6° per min and at 360° for a further 10 min. (Hamirin Kifli, Ph.D. thesis 1981, University of St. Andrews.)

0-6 double bonds. Table 3.4 includes results obtained by silver ion chromatography followed by gas chromatography of the methyl esters derived from each glyceride fraction.

TABLE 3.4. *Component Glycerides of Some Vegetable Oils by Silver Ion Chromatography*[a]

	322[b]	321	320	222	221	220	211	210	200	111	110
Soybean[c]	7	5	4	15	16	13	8	12	4	2	5
Peanut[d]	—	—	—	—	5	—	23	14	4	26	23

[a]F. D. Gunstone and M. I. Qureshi, *J. Amer. Oil Chemists' Soc.* 1965, **42,** 957, 961.
[b]The symbol 322 refers to all the triacylglycerols having one linolenic (3) and two linoleic (2) chains. The other three-figure arrays are to be interpreted similarly.
[c]Also 332, 1; 330, 1; 311, 2; 310, 3; 300, 2.
[d]Also 100, 5.

Glycerides with up to ~90 carbon atoms can be separated by gas chromatography on a short column (15 cm upwards) coated at a low level (1–3 %) with a suitable non-polar stationary phase, usually operated with temperature-programmed elution. Unsaturated glycerides are not separated from their saturated analogues under these conditions, but it is not difficult to separate glycerides differing by two carbon atoms. Some new liquid phases with increased polarity and moderate temperature stability permit the resolution of neutral lipid mixtures according to molecular weight *and* degree of unsaturation. Table 3.5 lists results available from a combination of silver ion chromatography and gas chromatography for triacylglycerols present in mutton fat.

TABLE 3.5. *Distribution by Carbon Number*[a] *of Triacylglycerols of Differing Unsaturation in Sheep Depot Fat*[b]

Carbon number[a]	Saturate	Monoene	Diene	Triene	Tetraene	Total
44	1.7	0.3	—	—	—	0.4
46	6.7	1.3	0.2	0.3	1.0	2.3
48	17.6	6.5	1.3	2.3	2.9	7.5
50	29.5	21.3	8.7	7.8	11.2	18.5
52	29.8	38.7	41.3	30.5	33.1	37.5
54	14.3	31.1	47.9	57.7	49.9	32.9
56	0.2	0.6	0.6	1.4	1.9	0.8
Total	31.3	45.8	19.4	2.6	0.8	100.0

[a]The carbon number is the total number of carbon atoms in the three acyl chains and does not include the three glycerol carbon atoms.
[b]Taken from F. D. Gunstone, *Comprehensive Organic Chemistry* (ed. Sir Derek Barton and W. D. Ollis), Pergamon 1979, p. 647.

These techniques can also be applied satisfactorily to wax esters and to sterol esters, to mono- and diacylglycerols after suitable derivatisation, and to phosphoglycerides after appropriate modification. Partial glycerides are derivatised to reduce polarity and to increase stability before chromatography. Acetates are preferred for thin-layer chromatography and trimethylsilyl ethers for gas chromatography.

Before being analysed by the separation procedures under consideration, phosphoglycerides are converted to diacylglycerols (which are then examined as described above) or to dimethylphosphatidic acids by the procedures outlined in Scheme 3.2. Some illustrative results are given in Table 3.6.

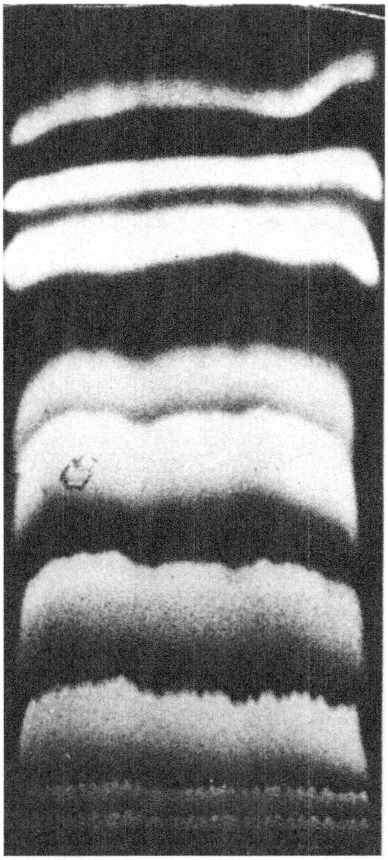

FIG. 3.2. Silver ion chromatography of soybean oil. 001 refers to triacylglycerols with two saturated acyl chains and one monounsaturated chain. The other numbers are interpreted similarly. (Adapted from Carter Litchfield, *Analysis of Triglycerides*, Academic Press, New York, 1972, p. 59, and used with permission.)

TABLE 3.6. *Comparison of the Major Phosphatidylcholines and Phosphatidylethanolamines of Fat Liver*[a]

Component acids C-1	C-2	PC	PE
16:0	18:1	6.4	0.9
16:0	18:2	15.7	7.2
18:0	18:2	9.0	5.3
16:0	20:4	18.9	14.2
18:0	20:4	20.9	35.0
18:1	20:4	3.6	4.1
16:0	22:6	4.8	12.4
18:0	22:6	5.6	8.8

[a]Taken from F. D. Gunstone, *Comprehensive Organic Chemistry* (ed. Sir Derek Barton and W. D. Ollis), Pergamon 1979, p. 648.

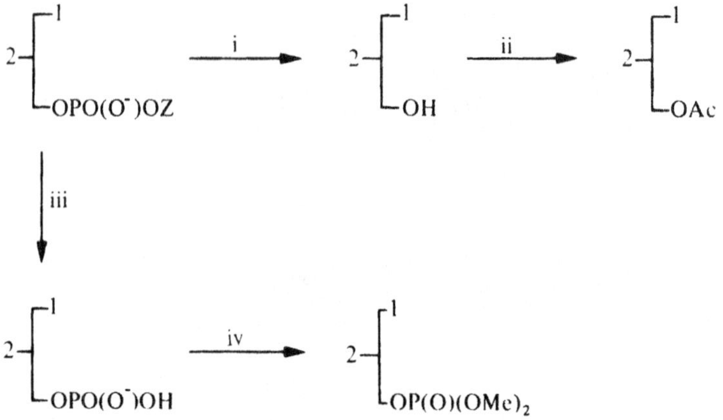

Modification of phosphoglycerides for further separation by thin-layer or gas chromatography (Z = choline, ethanolamine, etc.)
i, Phospholipase C; ii, Ac_2O, pyridine; iii, phospholipase D; iv, CH_2N_2.

SCHEME 3.2

REFERENCES

1. R. G. ACKMAN, *Progress in the Chemistry of Fats and Other Lipids*, 1972, **12,** 165.
2. A. KUKSIS, *Handbook of Lipid Research, Vol. 1, Fatty Acids and Glycerides*, (ed. A. Kuksis), Plenum Press, New York, 1978, p. 1.
3. F. D. GUNSTONE, *Recent Advances in the Chemistry and Biochemistry of Plant Lipids* (ed. T. Galliard and E. I. Mercer), Academic Press, London, 1974, p. 21.
4. G. R. KHAN and F. SCHEINMANN, *Progress in the Chemistry of Fats and Other Lipids*, 1979, **15**, 343.
5. A. KUKSIS, *Progress in the Chemistry of Fats and Other Lipids*, 1972, **12,** 1.
6. J. J. MYHER, *Handbook of Lipid Research, Vol. 1, Fatty Acids and Glycerides* (ed. A. Kuksis), Plenum Press, New York, 1978, p. 123.
7. K. AITZETMÜLLER, *J. Chromatog.* 1975, **113,** 1231.
8. W. C. BRECKENRIDGE, *Handbook of Lipid Research, Vol. 1, Fatty Acids and Glycerides* (ed. A. Kuksis), Plenum Press, New York, 1978, p. 197.
9. D. BUCHNEA, *Handbook of Lipid Research, Vol. 1, Fatty Acids and Glycerides* (ed. A. Kuksis), Plenum Press, New York, 1978, p. 233.

GENERAL REFERENCE

CHRISTIE, W. W. *Lipid Analysis*, 2nd ed. 1982, Pergamon, Oxford.

CHAPTER 4

The Biosynthesis and Metabolism of Fatty Acids and Lipids

The biosynthesis and metabolism of fatty acids in plant and animal systems have been studied intensively and the basic pathways are now known. In this brief account of the subject attention is directed to the chemistry of the intermediates and of the reactions rather than to the nature of the enzymes or the control mechanisms. In some of the reactions to be discussed the true nature of the lipid substrate is not fully known and reference to compounds such as acetate or palmitate refer to the acid combined in an appropriate form. The acid may be in a free form, it may be bound to an enzyme or a coenzyme, or it may already be incorporated into a lipid. The acids are commonly present as thiol esters and representations such as RCOSCoA or RCOSACP signify the acid RCOOH attached to a thiol (SH) group present in coenzyme A (CoASH) or acyl carrier protein (ACPSH) respectively.

A. FATTY ACID BIOSYNTHESIS

1. *de novo* Synthesis of palmitic and other saturated acids

The first stage of fatty acid biosynthesis is usually the *de novo* production of palmitic (or other saturated) acid from acetate. Each molecule of palmitic acid (16:0) is derived from eight molecules of acetate (2:0), but not all the acetate molecules behave in the same way. One acetate molecule enters the condensation process directly as the starter or primer; the remaining seven are first converted to malonate ($HOOCCH_2COOH$). The additional carbon atom in each malonate molecule is, however, lost during condensation so that all sixteen palmitic carbon atoms are acetate-derived.

starter: acetate
chain - extender: 7 acetate ⟶ 7 malonate
} ⟶ palmitate + $7CO_2$

The conversion of acetate to malonate is catalysed by a biotin-containing enzyme (acetyl-CoA carboxylase) and the remainder of the chain-extension process is catalysed by fatty acid synthetase, a group of enzymes which exist in yeast and in animal tissues as a multi-enzyme complex or in bacterial and plant tissues in a readily dissociable form. The synthetase contains enzymes to catalyse each of the five steps in the cycle:

(i) a *transacylase* to transfer acetyl and malonyl groups from coenzyme A to appropriate thiol functions in acyl carrier protein.
(ii) a *condensing enzyme* to combine acetate and malonate to furnish 3-oxobutanoate (acetoacetate) with liberation of carbon dioxide (this contains the carbon atom used to convert acetate to malonate).
(iii) a *reductase* which, along with NADPH, gives stereospecifically 3 *R*-(−)-hydroxybutanoate,
(iv) a *dehydrase* giving *trans* but-2-enoate by a dehydration that is both regiospecific and stereospecific, and
(v) another *reductase* which with NADPH or NADH gives butanoate.

At this stage the first cycle is complete and acetate (C_2) has been converted to butanoate (C_4). Subsequently this reacts with other malonate molecules to give successively the C_6, C_8, C_{10}, C_{12}, C_{14}, and C_{16} acids. Finally, another enzyme hydrolyses the carboxylic acid from the fatty acid synthetase, thereby controlling the chain-length of the saturated fatty acid. The major product of most *de novo* systems is palmitic acid, but this is sometimes accompanied by stearic acid and there must be plant systems producing mainly lauric and myristic acids. This sequence of changes is set out in Scheme 4.1.

(i) formation of malonyl - CoA

$$ATP + HCO_3^- + \text{biotin - enzyme} \xrightleftharpoons{Mn^{2-}} ADP + P_i + CO_2\text{ - biotin-enzyme}$$

$$CO_2\text{-biotin-enzyme} + CH_3COSCoA \rightleftharpoons \text{biotin-enzyme} + HO_2CCH_2COSCoA$$

(ii) transfer to acyl carrier protein (ACPSH)

$$CH_3COSCoA + ACPSH \rightleftharpoons CH_3COSACP + CoASH$$

$$HO_2CCH_2COSCoA + ACPSH \rightleftharpoons HO_2CCH_2COSACP + CoASH$$

(iii) condensation

$$CH_3COSACP + HO_2CCH_2COSACP \rightleftharpoons CH_3COCH_2COSACP + CO_2 + ACPSH$$

(iv) reduction

$$CH_3COCH_2COSACP + NADPH + H^+ \rightleftharpoons CH_3CH(OH)CH_2COSACP + NADP$$

(v) dehydration

$$CH_3CH(OH)CH_2COSACP \rightleftharpoons CH_3CH \overset{t}{=} CHCOSACP + H_2O$$

(vi) hydrogenation

$$CH_3CH{=}CHCOSACP + NADPH + H^+ \rightleftharpoons CH_3CH_2CH_2COSACP + NADP$$

SCHEME 4.1. Biosynthesis of saturated acids by the *de novo* pathway.

This common biosynthetic pathway also occurs in a number of modified forms. One uncommon variant leads to monoenoic acids (Section 4.3), though this is not the usual route to such acids. In another variation the starter or primer acetic acid molecule (but not the extender acetate units which react as malonate) is replaced by a different compound. Chain-extension by C_2 units follow the normal *de novo* pathway and the final product is a homologous series of acids with one member predominating. Propionate (C_3) gives "odd" acids (mainly heptadeconoate), 2-methylpropionate gives iso acids with an even number of carbon atoms per molecule (mainly 16-

methylheptadecanoate), and 2-methylbutanoate gives anteiso acids with an odd number of carbon atoms per molecule (mainly 14-methylhexadecanoate).

In another modification one or more of the acetate (or malonate) units in the chain-extension process are replaced by propionate (or methylmalonate) units. A methyl branch is introduced each time one of these units is incorporated into what would otherwise be an acetate–malonate condensation process. For example, branched-chain acids produced by lambs fed on a barley-rich diet arise in this way.

2. Chain elongation

A slightly different elongation process proceeding through similar intermediates converts the acid RCOOH to its bishomologue RCH_2CH_2COOH and occurs with saturated and unsaturated substrates. Acetate or malonate may be involved in the chain-extension process and the whole reaction cycle occurs via coenzyme A derivatives. A typical example is the chain-extension of oleate to 20:1 and 22:1 (erucate) in two elongation cycles:

$$18{:}1\ (9c) \longrightarrow 20{:}1\ (11c) \longrightarrow 22{:}1\ (13c)$$

The extender acetate or malonate may be replaced by propionate or methylmalonate to furnish polymethyl branched acids, e.g.

$$20{:}0 \longrightarrow 2\text{-methyl } 22{:}0 \longrightarrow 2{,}4\text{-dimethyl } 24{:}0 \text{ etc.}$$

3. Monoene biosynthesis

Monoene acids are produced in two ways. The less common route, confined to lower forms of life having an anaerobic existence, is not discussed here.

The more important route to monoenes occurs in plants, animals, and micro-organisms. Oxygen and NADH (or NADPH) are also required. A double bond of *cis* configuration is produced by removal of the pro-*R* hydrogen atoms from C-9 and C-10 and the substrate may be a coenzyme A ester, an ACP ester, or an acyl chain already incorporated into a phospholipid. The desaturation produces Δ9 acids of varying chain-length which may be converted to other monoene acids by chain-elongation (Section 2).

4. Polyene biosynthesis

Polyene acids are important as precursors of prostaglandins and as constituents of membrane lipids (Section E). Plants convert monoenes to polyenes by further desaturation of the distal end of the molecule (between the existing double bond and the ω-methyl group), additional *cis*-unsaturated centres being introduced in a methylene-interrupted pattern. In contrast, animals insert additional double bonds in the proximal end (between the existing double bond and the carboxyl group) of monoenes and of plant-derived polyenes, but never in the distal unit (Scheme 4.2).

	Distal	Proximal
	$CH_3CH_2CH_2CH_2CH_2CH_2CH_2CH_2CH = CHCH_2CH_2CH_2CH_2CH_2CH_2CH_2CO_2H$	
plants:	commonly	rarely
animals	never	commonly

SCHEME 4.2.

oleic 18:1 (9*c*) —animals→ *n* - 9 polyenes

↓ plants

linoleic 18:2 (9*c*12*c*) —animals→ *n* - 6 polyenes

↓ plants

linolenic 18:3 (9*c*12*c*15*c*) —animals→ *n* - 3 polyenes

SCHEME 4.3

Plants convert oleate to linoleate and linolenate and these three unsaturated C_{18} acids are the most common in the plant kingdom. Each serves as precursor for a family of polyene acids (Scheme 4.3 and Table 1.4).

Animals, unable to insert double bonds in the distal end of the molecule, produce polyene acids by desaturation and chain-elongation of oleate (*n*-9-acids), hexadec-9-enoate (*n*-7 acids), linoleate (*n*-6 acids), and linolenate (*n*-3 acids). The two monoene precursors may be endogenous or exogenous in origin, but the linoleate and linolenate must come from plants, since animals cannot biosynthesise these acids.

In addition to the 9-desaturase, which serves mainly to convert saturated acids to Δ9 monoenes, there are only three other common desaturases (6-, 5-, and 4-). The 6-desaturase operates most effectively on substrates having a 9*c* double bond (with or without additional unsaturated centres). Similarly the best substrates for the 5- and 4-desaturases have 8*c* and 7*c* unsaturation respectively. The formation of the *n*-6 acids and the *n*-3 acids is set out in Schemes 4.4 and 4.5 respectively. Acids on the same horizontal line are linked by elongation and chain-shortening processes. Passage to a lower horizontal line occurs through the substrate most appropriate for the 6-, 5- and 4-desaturase respectively. Linoleic acid is converted most often to arachidonic acid (20:4), whilst linolenic acid furnishes mainly icosapentaenoic and docosahexaenoic acids.

5. Essential fatty acids and prostaglandins

Essential fatty acid deficiency in humans results in abnormal skin conditions such as scaliness and dermatitis, increased water loss, reduced regeneration of tissues, increased susceptibility to infection, and an increase in the ratio of 20:3 (*n*-9) to 20:4 (*n*-6) acids in the serum phospholipids. These effects are probably due, at least in part, to the lack of prostaglandins produced from specific C_{20} polyene acids.

Though difficult to induce in man, essential fatty acid deficiency is observed in certain diseased states and Holman claims that this deficiency, far from being of no consequence, is of frequent

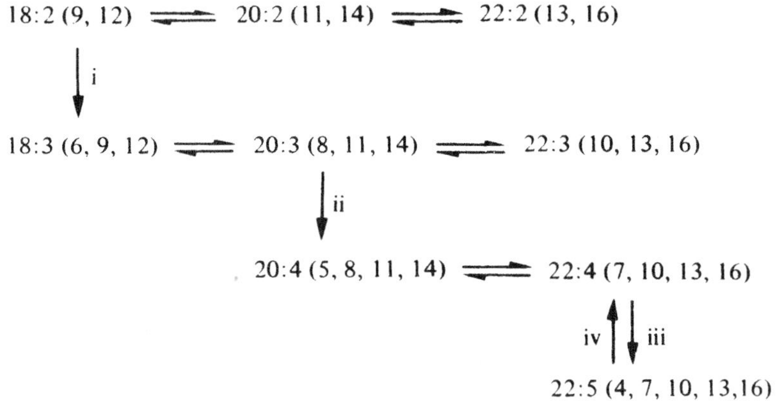

i, 6-desaturase; ii, 5-desaturase; iii, 4-desaturase; iv, 4-enoylreductase.

SCHEME 4.4. *n*-6 Polyene acids.

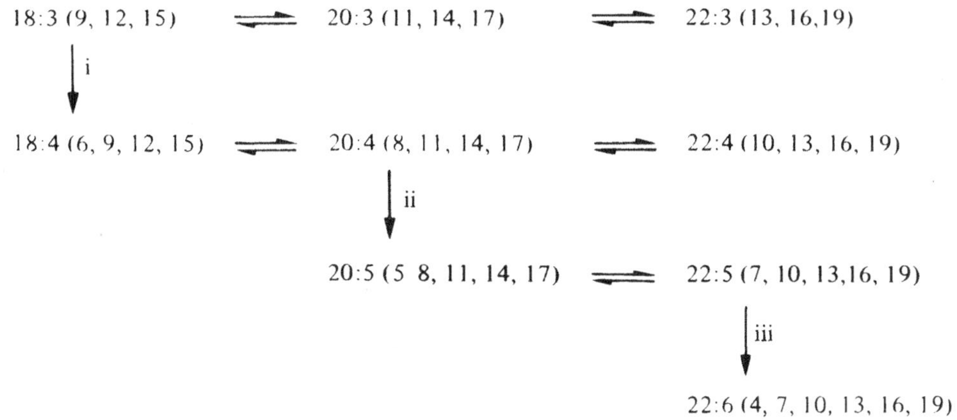

i, 6-desaturase; ii, 5-desaturase; iii, 4-desaturase.

SCHEME 4.5. *n*-3 Polyene acids.

occurrence. It is often induced in hospital patients subject to long-term intravenous feeding with fat-free preparations and occurs in genetic diseases in which essential fatty acids are not properly utilised because of enzymic deficiencies. Such a deficiency may be enhanced or promoted by abnormal amounts of other dietary constituents. No large storage depots of polyene acids are known in animals, and these must therefore be obtained from vital metabolic membranes with the possibility of undesirable consequences for general metabolism.

EFA deficiency can usually be cured by linoleic acid because this acid—which cannot be produced in animals and is entirely of plant origin—can be metabolised to the required C_{20} and C_{22} polyene acids (Scheme 4.4). Such acids are preferentially incorporated into cell membrane phospholipids where they have an important structural and functional role.

EFA status is best assessed biochemically by the ratio of the 20:3 (*n*-9) acid produced from oleic to the 20:4 (*n*-6) acid produced from linoleic acid in the serum phospholipids. Values below 0.4 have generally been regarded as acceptable, but recent studies indicate that the 0.3 is a safer limit.

The recommended daily intake of EFA for adult humans has been quoted as at least 3% of energy requirement or as 2–10 g per day of linoleic acid.

Natural prostaglandins are produced in many animal tissues from the 20:3, 20:4, and 20:5 polyene acids which furnish the PG_1, PG_2, and PG_3 series of compounds respectively. The several members of each series vary in the degree of oxygenation. All prostaglandins are derivatives of the cyclic C_{20} acid, prostanoic. These compounds are not stored in mamalian tissue, but are elaborated in response to various stimuli. Virtually every mammalian tissue examined has some capacity to synthesise PG related material.

Very unusually the Carribean coral *Plexaura homoalla* contains up to 3% of its dry weight as PGA_2 in the form of the 15-acetate methyl ester and its 15-epimer.

The enzymic production of prostaglandins is inhibited by the acetylenic analogue of arachidonic acid (20:4 5*a*8*a*11*a*14*a*) and also by aspirin, paracetamol, and indomethicin. The prostaglandins

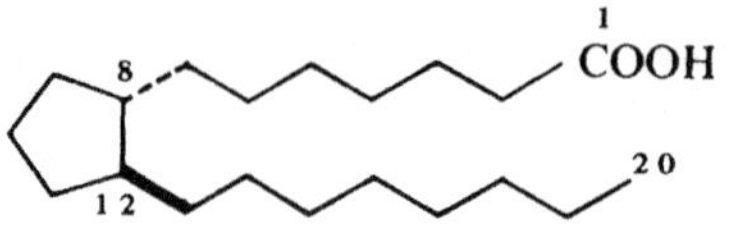

prostanoic acid

PG_1 compounds with one double bond (13*t*) are derived from 20:3 (8, 11, 14)

PG_2 compounds with two double bonds (5*c*, 13*t*) are derived from 20:4 (5, 8, 11, 14). Some examples are shown below.

PG_3 compounds with three double bonds (5*c*,13*t*, 17*c*) are derived from 20:5 (5, 8, 11, 14, 17)

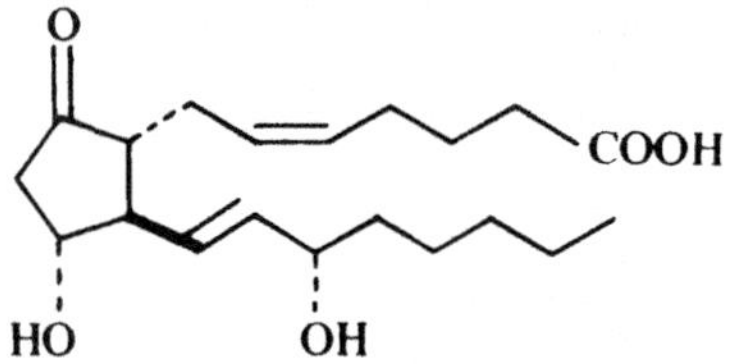

PGE_2

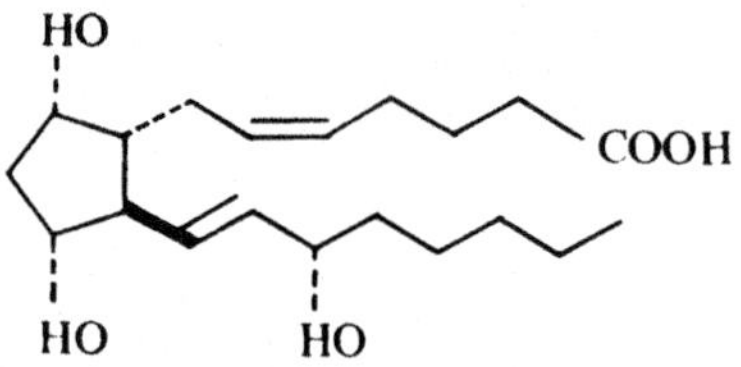

PGF_{2a}

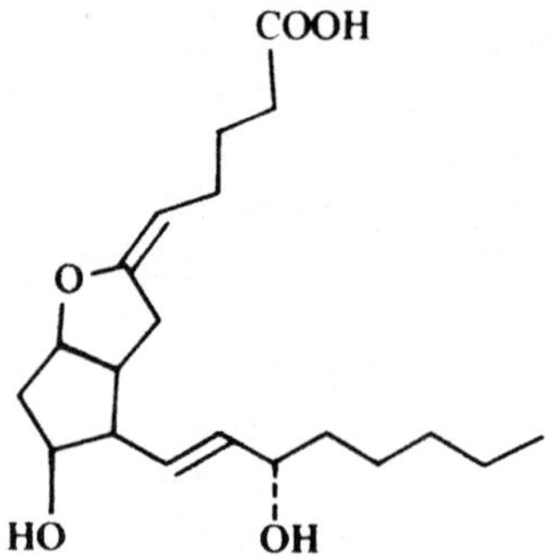

PGI_2

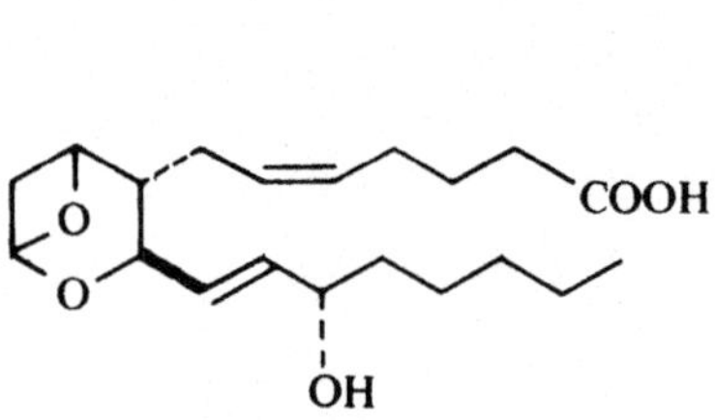

TXA_2

SCHEME 4.6. Structure of prostaglandins and thromboxanes.

are quickly metabolised by (a) oxidation of the 15-hydroxy group to 15-oxo by prostaglandin-dehydrogenase, (b) reduction of the *trans* 13 double bond by 13,14-reductase, (c) ω-oxidation, and (d) β-oxidation of the carboxyl chain and of the methyl chain after ω-oxidation. The primary deactivation step is usually the oxidation of the 15-hydroxy function. For example, more than 95% of PGF_2 present in the blood stream is deactivated by prostaglandin-dehydrogenase in a single passage through the lungs.

Prostaglandins are used in childbirth to induce parturition. In larger doses they can induce abortion. They may also be used for the treatment of gastric ulcers and bronchial asthma and they inhibit platelet aggregation. They play a role as luteolytic substances and are used in the veterinary field for the control of oestrus. Each prostaglandin is specific in its physiological effects and different prostaglandins sometimes show opposing effects.

In the presence of microsomes from the rabbit or from pig aorta the prostaglandin endoperoxides (PGG_2, PGH_2) are converted to prostacyclin (PGI_2, formerly PGX). This compound inhibits platelet aggregation and may even bring about its reversal. In the musculature there exists the remarkable possibility that the endoperoxides produced by platelets are converted by platelets into potent vasoconstricting and platelet-aggregating thromboxanes and by the vessel walls into potent vasodilating and platelet-aggregate-inhibiting prostacyclin. The balance between the antagonistic properties of thromboxane and prostacyclin may maintain blood vessel tone and platelet functionality and be critical for thrombus formation.

In the presence of washed human platelets arachidonic acid gives some open-chain hydroxy acids and a cyclic compound, thromboxane B_2, formed via the less-stable thromboxane A_2. The latter is found widely in the body, but has a half-life of only 36 s in water.

B. LIPID BIOSYNTHESIS

1. General comments

Only a simplified account of lipid biosynthesis is presented here. Before discussing the more important lipid classes some general points are made:

(i) The glycerol present in glycerolipids is derived from glycerol, glyceraldehyde (2,3-dihydroxypropanal), or dihydroxyacetone (1,3-dihydroxypropan-2-one), each of which is a product of carbohydrate metabolism.
(ii) Most lipids can be produced by more than one pathway.
(iii) Apart from the ether lipids, all glycerolipids are made from phosphatidic acids or 1,2-diacylglycerols which are themselves interconvertible (Scheme 4.7).
(iv) Structural units to be attached to the phosphatidic acids or 1,2-diacylglycerols are made available in a modified form, usually associated with a nucleotide such as cytidine or uridine phosphate.
(v) Acylation is usually effected by fatty acids as their coenzyme A thiol esters.

2. Triacylglycerols

The major *de novo* route to triacylglycerols from glycerol (or glyceraldehyde or dihydroxyacetone) proceeds via a phosphatidic acid and a 1,2-diacylglycerol with each step under

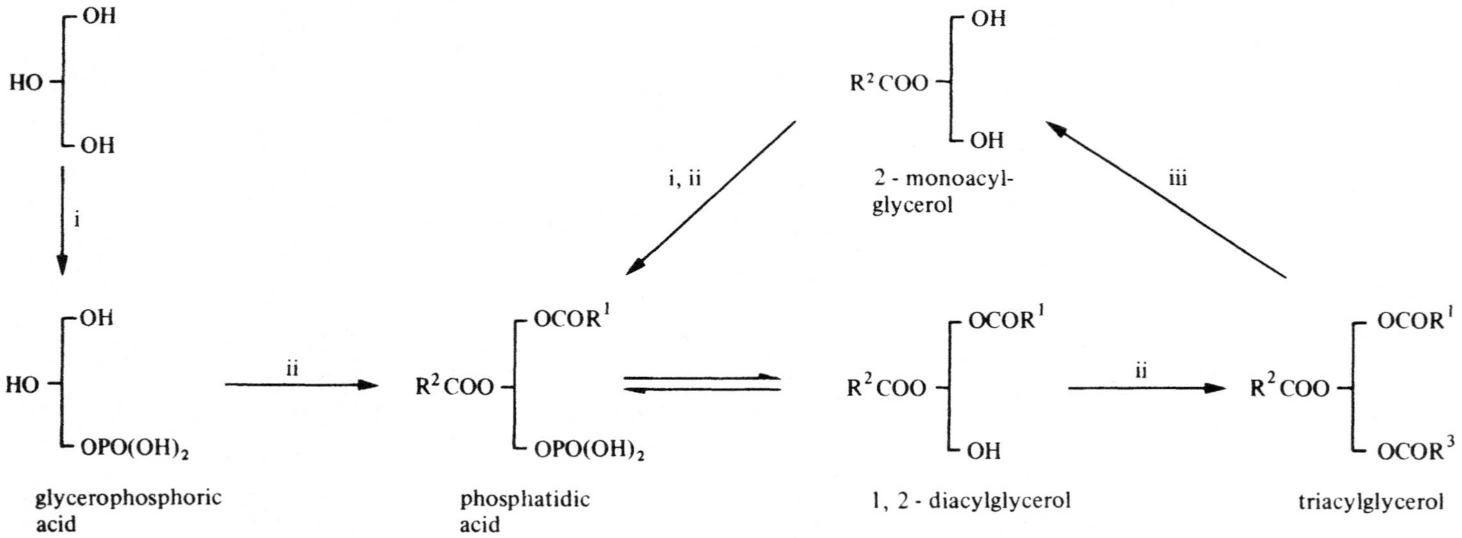

i, ATP; ii, acylation; iii, lipolysis

SCHEME 4.7. Biosynthesis of triacylglycerols.

the influence of a different enzyme. The alternative process of reacylation of 2-monoacylglycerols, resulting from lipolysis of triacylglycerol, is significant in animals on a fat-containing diet.

3. Phosphoglycerides

Phosphatidylcholine (ethanolamine) results from interaction of 1,2-diacylglycerol with cytidine monophosphate phosphocholine (ethanolamine). The phosphatidylethanolamines can also be converted to phosphatidylcholines by progressive methylation by S-adenosylmethionine.

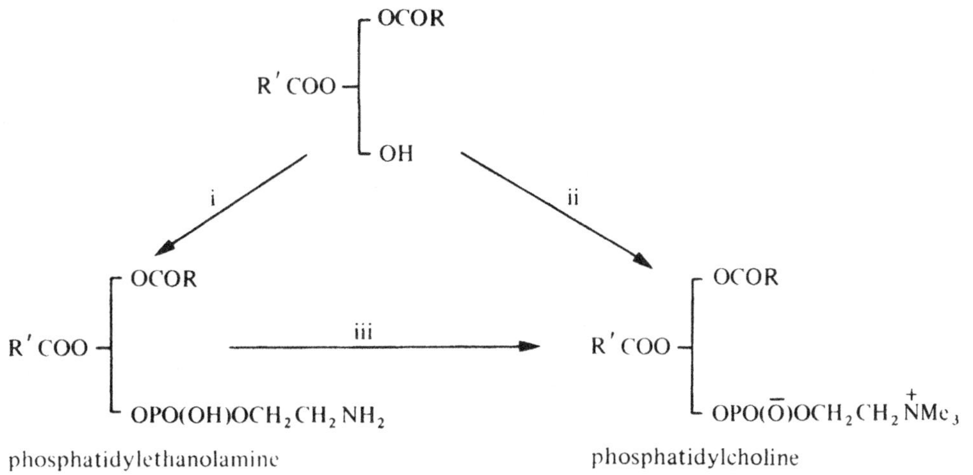

i, CMPethanolamine; ii, CMPcholine; iii, S-adenosylmethionine.

SCHEME 4.8. Biosynthesis of phosphatidylethanolamine and phosphatidylcholine.

C. FATTY ACID AND LIPID METABOLISM

1. The digestion and absorption of fats

Dietary fats of animal or vegetable origin are classified as visible (adipose tissue, milk fat, seed oils) or invisible (derived from animal or vegetable membranes). They are the richest source of energy on a weight basis and excess of fat beyond daily energy requirement is laid down as reserve depot fat (possibly after modification). But dietary fats also serve other purposes. They play a structural role in membranes which may have requirements for specific fatty acids, they affect the texture of food and generally make it more palatable, they serve as a solvent for the fat-soluble vitamins (A, D, E), and they contribute to food flavour in both desirable and undesirable ways.

In the total picture of food intake it should be noted that (a) blood glucose must be maintained within very narrow limits since the brain cannot use free fatty acid as food—only glucose, (b) excess of carbohydrate is stored first as glycogen mainly in the liver and muscle and then as fat in adipose tissue, (c) protein supply must be at an adequate level, but excess of protein is converted to fat, and (d) when energy is required it is obtained first by mobilisation of glycogen reserves, then from fat, and finally from protein.

Most adults ingest 120–150 g of dietary fat per day. This is mainly triacylglycerol and in normal health is absorbed almost completely. Fat is converted in the stomach to a course emulsion and then passes into the duedenum where, mixed with bile, pancreatic juice, and chyme, it is partially hydrolysed to free acids, 2-monoacylglycerols, and some glycerol. Most of the glycerol, along with short-chain acids ($<$C10), passes into the portal blood. The monoacylglycerols and free acids enter the intestinal mucosa where triacylglycerols are regenerated and combined with proteins in chylomicrons. These pass into general circulation *via* the thoracic duct and are transported in the blood stream to the liver ($\sim$30%), fat depots ($\sim$30%), and the musculature and other organs ($\sim$40%). The blood stream also contains triacylglycerols and free acids (as albumin complexes) coming from the liver and the fat depots.

During fasting, free acids from the fat depots circulating as albumin complexes are removed by peripheral tissues where they are oxidised or by the liver ($\sim$40%) where some are oxidised and some converted to triacylglycerols before being returned to the blood.

2. Bio-oxidation of fatty acids

Bio-oxidation of fatty acids serves two different purposes. Most commonly, complete oxidation furnishes energy required for other biological purposes; less commonly, partial oxidation produces other acids required for a particular purpose. Energy production is achieved mainly by β-oxidation and sometimes by ω-oxidation. Fatty acid modification results from α- or ω-oxidation and occasionally by β-oxidation.

Fatty acids contain more stored energy per carbon atom than any other biological fuel. They are degraded by enzymic processes to carbon dioxide and water with most of the energy produced being stored as ATP.

β-Oxidation. β-Oxidation is the most common mechanism of degradation. It involves a cycle of reactions whereby a fatty acid or its CoA derivative gives acetyl-CoA along with an acyl-CoA having two less carbon atoms than the original acid. The latter enters again into the reaction cycle. Acetyl-CoA interacts with oxaloacetate and is converted via the citric acid cycle to water and carbon dioxide. The β-oxidation cycle proceeds through the following intermediates:

$$RCH_2CH_2COOH \xrightarrow{\text{i}} RCH_2CH_2COSCoA \xrightarrow{\text{ii}} RCH{=}CHCOSCoA$$

$$\xrightarrow{\text{iii}} RCH(OH)CH_2COSCoA \xrightarrow{\text{iv}} RCOCH_2COSCoA$$

$$\xrightarrow{\text{v}} RCOSCoA + CH_3COSCoA$$

In step (i) the fatty acid is converted to its CoA ester through reaction with ATP and coenzyme A (CoASH) in the presence of Mg^{2+} and an acyl-CoA synthetase. This ester is first produced outside the mitochondrion, but β-oxidation occurs within the mitochondrial matrix and passage through the membrane actually occurs as an acyl derivative of L-carnitine ($\overset{-}{O}OCCH_2CH(OH)CH_2\overset{+}{N}Me_3$).

The CoA ester is then regenerated within the mitochondrion. This is dehydrogenated to the $\Delta 2t$ dehydroacyl-CoA with acyl-CoA dehydrogenase (step ii) and hydrated (step iii) in a regiospecific and stereospecific reaction occurring under the influence of enoyl-CoA hydratase. The $\Delta 2t$ ester gives the 3L(+)-hydroxyl derivative and, less commonly, the 2*c* ester gives the 3D-(−)-hydroxy ester. Step (iv) is also stereospecific in that the oxidising enzyme (L-hydroxyacyl-CoA-dehydrogenase), which requires NAD^+, is specific for the 3-L-(+)-enantiomer. Finally, acetyl-acyl-CoA transacylase and coenzyme A are required to effect cleavage (step v).

n-Saturated acids with an even number of carbon atoms are degraded by *β*-oxidation to acetyl-CoA only whilst *n*-saturated acids with an odd number of carbon atoms give acetyl-CoA (several molecules) and propionyl-CoA (one mole). This is further degraded via the C_4 compounds methylmalonate and succinate. This last compound enters the citric acid cycle.

The fate of acids with one or more branched methyl groups depends on whether these are attached to even or odd carbon numbers. On an even carbon atom the 2-methyl-branched acid produced at some stage during the oxidation furnishes propionyl-CoA which is further degraded as indicated above. On an odd carbon atom a 3-methyl-branched acid is produced and acetoacetate and 3-hydroxy-3-methylglutarate are formed as intermediate metabolic products.

Other problems arise with the common unsaturated acids which at some stage in the *β*-oxidation furnish Δ2*c* or Δ3*c* acids. Oleic acid, for example, after three oxidation cycles gives 12:1 (3*c*). This is isomerised to the 2*t* acid and *β*-oxidation can then continue in the normal way. With linoleic acid the 9*c* double bond is handled as in oleic acid. The 12*c* double bond eventually appears as 8:1 (2*c*) which is hydrated to the D-(−)-3-hydroxy acid. This can be dehydrogenated only after epimerisation to the L-(+)-enantiomer. This isomerase and epimerase must therefore be included in the group of enzymes required for *β*-oxidation. Polyene acids with Δ4 unsaturation may undergo hydrogenation at this position before *β*-oxidation commences.

18:2 (9*c*12*c*) ---→ 16:2 (7*c*10*c*) ---→ 14:2 (5*c*8*c*) ---→ 12:2 (3*c*6*c*)

⟶ 12:2 (2*t*6*c*) ---→ 10:1 (4*c*) ---→ 8:1 (2*c*) ⟶ 3D - OH 8:0

⟶ 3L - OH 8:0 ---→ 6:0 ---→ 4:0 ---→ 2:0

Occasionally *β*-oxidation is not complete and lower homologues resulting from incomplete oxidation are found in a lipid sample. For example, hexadec-7-enoic acid and hexadeca-7,10-dienoic acid probably result from oleic acid and linoleic acid respectively by a single *β*-oxidation cycle.

α-Oxidation. α-Oxidation which can convert a fatty acid to its lower homologue or to the α-hydroxy derivative is now considered to be a reaction of the free acid occurring by a radical process via the α-hydroperoxy acid thus:

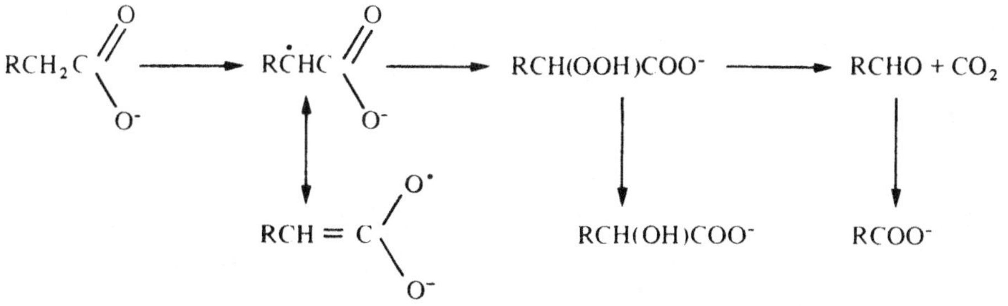

ω-Oxidation. ω-Oxidation produces an ω or ω-1-hydroxy acid through the influence of a hydroxylase and these compounds can be oxidised further:

$$CH_3(CH_2)_{n+1}COOH \rightarrow HOCH_2(CH_2)_{n+1}COOH \rightarrow OHC(CH_2)_{n+1}COOH \rightarrow HOOC(CH_2)_{n+1}COOH$$

$$CH_3(CH_2)_{n+1}COOH \rightarrow CH_3CH(OH)(CH_2)_nCOOH \rightarrow CH_3CO(CH_2)_nCOOH$$

This process is commonly observed in compounds in which β-oxidation has been blocked as in the conversion of 2,2-dimethyloctadecanoic acid to 2,2-dimethylhexanedioic acid in which ω-oxidation is followed by repeated β-oxidation.

$$CH_3(CH_2)_{15}C(CH_3)_2COOH \xrightarrow{\omega\text{-oxidation}} HOOC(CH_2)_{15}C(CH_3)_2COOH \xrightarrow[\beta\text{-oxidation}]{\text{repeated}} HOOC(CH_2)_3C(CH_3)_2COOH$$

D. POSSIBLE HARMFUL EFFECTS OF SOME DIETARY LIPIDS

1. Heart disease

Simplistically heart disease can be related to the deposition of serum lipids in the interior of blood vessels. The resulting scleroma build up until the vessel becomes severely occluded and lacks pliability (hardening of the arteries). A blood clot may then occur in the brain (a stroke) or in the heart (heart attack). The clot contains mucopolysaccharide, phospholipid, and cholesterol especially as ester. The diet generally recommended for the prevention of this condition (atherosclerosis) among population groups with a high incidence of atherosclerosis should contain energy sufficient to maintain ideal body weight with 10–15 % of the energy requirement as protein and 30–35 % as fat of which not more than one-third should be saturated acids and not less than one-third should be linoleic acid. Further, the diet should be low in refined sugar and alcohol and contain less than 300 mg of cholesterol per day (WHO report). It is considered that such a diet will significantly decrease the two main risk factors for atherosclerosis, *viz.* blood lipoprotein containing cholesterol and triacylglycerols and the thrombotic tendency of blood platelets.

2. Long-chain monoene acids

A problem has been noted with those rapeseed oils which have a high content of erucic acid (22:1, 20–55 %) along with smaller amounts of icosenoic acid (20:1, 10 %), though partially hydrogenated fish oils containing several 22:1 isomers may behave similarly. It is reported that diets containing appreciable amounts of erucic acid retard the growth of certain experimental animals; adversely affect various organs morphologically, biochemically, and functionally; lead to intracellular fat accumulation in heart and red skeletal muscle; and eventually to myocardial changes characterised by cellular infiltration, death of heart muscle, and the replacement of this by scar tissue. Although digestion and transport of erucic acid in humans is normal, there is evidence that this acid is less efficiently oxidised than the shorter-chain acids. The consequences are diminished with the low-erucic acid rapeseed oils which have been developed and can be eliminated by appropriate dilution with other oils. It is therefore concluded that the best precautions are those of moderation and avoidance of over-dependency on vegetable oil(s) of this type.

E. MEMBRANES

Biological membranes serve several useful purposes within a cell. In particular: they divide the system into compartments and function as the boundary structure of the cell, its organelles, and compartments; they are semi-permeable and regulate the passage of ionic and non-ionic substances; they contain proteins (enzymes) catalysing specific reactions; and they provide reception and other binding sites to interact with hormones and antibodies.

Membranes consist mainly of lipids and proteins along with small amounts of carbohydrates and other compounds. The proportion of lipid varies between 25 % and 75 %, but is usually around 50 %. The lipids vary in their nature so it is difficult to make more than general statements about their composition. They are mainly phospholipids (especially phosphatidylcholine, ethanolamine and -serine, cardiolipin, and sphingomyelin) and cholesterol. Glycolipids are sometimes present, but cholesterol esters and triacylglycerols are virtually absent from membranes.

Ideas about the structure of membranes have developed from the lipid bilayer model to the lipid-globular protein mosaic model (Fig. 4.1). A major problem is how the proteins interact with the lipids. The lipid is believed to assume an essentially bilayer structure with hydrophilic head groups on the outside in contact with an aqueous medium and hydrophobic fatty acid chains inside and not in contact with the aqueous medium. How are proteins and cholesterol incorporated into this bilayer structure? An early proposal suggested that a monolayer of globular protein was associated with the polar head groups on each side of the lipid bilayer, but the current view is that the regular nature of the lipid bilayer is disrupted to incorporate protein molecules. Some of these are exposed only at one side of the bilayer, whilst others extend through the bilayer and protrude from both sides. The membrane is thus conceived as a dispersion of globular proteins in a fluid lipid matrix. The fluidity of the membrane and the lateral mobility of its lipid and protein components are important in relation to its function.

The lipid components of biological membranes are distributed asymmetrically with respect to the inner and outer layers of the membrane. Phosphatidylcholines concentrate in the outer layer and glycolipids also occur on the outside of membranes with their attached sugar groups projecting into the surrounding water. The lipid composition of a membrane varies not only in respect of its lipids but also in its fatty acids. It is believed that in all organisms the membrane lipids

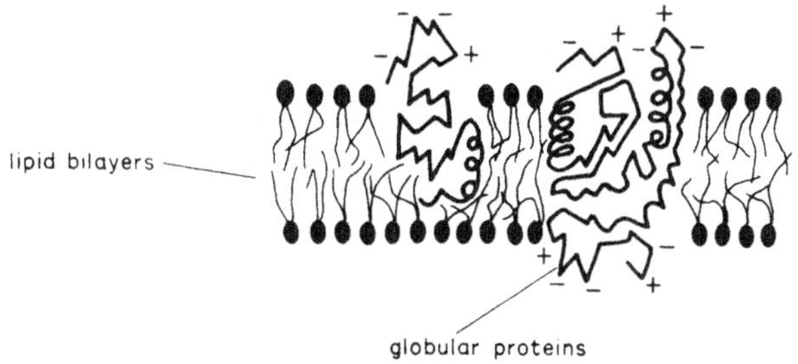

FIG. 4.1. The lipid-globular protein mosaic model. (Copied with permission from *Lipid Biochemistry—An Introduction*, by M. I. Gurr and A. T. James, 3rd ed. Chapman & Hall, London, 1980, p. 203.)

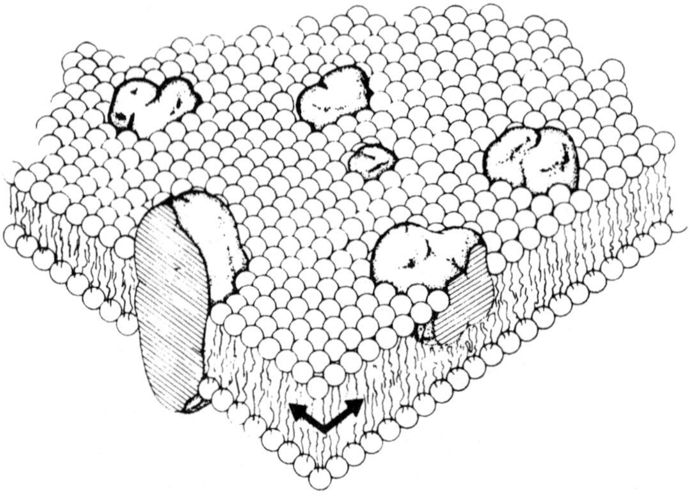

FIG. 4.2. The hydrophobic carbon chains of the lipids are represented as wavy lines and their hydrophilic head groups as spheres. Proteins are embedded in the membrane. (S. J. Singer and G. L. Nicolson, *Science*, 1972, **175**, 723.)

must be partially liquid at temperatures suitable for life processes. This is achieved by modification of the fatty acid composition: in particular, by increasing *cis*- unsaturation or by the presence of branched methyl groups or cyclopropane groups.

The alkyl chains of the fatty acids can assume many conformations and movement from one conformation to another occurs in the liquid state. They are free to rotate and to undergo twisting whilst the bilayer still holds together. Melting of the acyl chains is first observed at the pre-transition temperature and is completed at the transition temperature (T). Between these two temperatures solid and liquid regions coexist in the lipid bilayer. This melting behaviour can be related to the fatty acids in the lipid and pretransition and transition temperatures have been measured for several pure phospholipids. For dipalmitoylphosphatidylcholine, for example, the transition temperature is 40.8 and the pretransition temperature 29.5°.

The phospholipid and protein units of a membrane can also exchange places with neighbouring molecules of like kind. This happens quite regularly and exchange of phospholipid units is estimated to occur at the rate of about $10^7 s^{-1}$.

Though relatively stable, membrane components are not chemically inert and the phospholipid units are subject to oxidative and hydrolytic changes.

GENERAL REFERENCES

GALLIARD, T., and E. I. MERCER, (eds.) *Recent Advances in the Chemistry and Biochemistry of Plant Lipids*, Academic Press, London, 1975.

GURR, M. I., and A. T. JAMES, *Lipid Biochemistry—An Introduction*. 3rd edition, Chapman & Hall, London, 1980.

KUNAU, W.-H., and R. T. HOLMAN (eds.) *Polyunsaturated Fatty Acids*, American Oil Chemists' Society, Champaign, 1977.

O'DOHERTY, P. J. A., *Handbook of Lipid Research, Vol. 1, Fatty Acids and Glycerides* (ed. A. Kuksis), Plenum Press, New York, 1978, p. 289.

CHAPTER 5

Physical Properties

A. POLYMORPHISM AND CRYSTAL STRUCTURE[1,2]*

1. Introduction

Many long-chain compounds exist in more than one crystalline form and may consequently exhibit more than one melting point. This property of polymorphism is of considerable academic and industrial interest for the understanding of polymorphic change is essential for the satisfactory blending and tempering of those fat-containing materials such as cooking fats and ice cream which must attain and maintain a certain physical appearance during preparation and storage. Problems of graininess in margarine and bloom in chocolate are related to polymorphic changes.

There are several experimental methods of examining polymorphism, but the most extensively employed are the study of melting behaviour through heating and cooling curves and dilatometry, infrared spectroscopy, and X-ray diffraction. It has, however, proved difficult to correlate the observations made in these different ways. The problem is complicated by the difficulty of obtaining pure compounds (especially glycerides) for basic studies and by the fact that natural fats are complex mixtures.

X-ray investigations indicate that the unit cell for long-chain compounds is a prism with two short spacings and one longer one (Fig. 5.1). When the long-spacing is shorter than that calculated from known bond lengths and angles, it is assumed that the molecule is tilted with respect to its end planes. On the other hand, the length may be such as to indicate a dimeric unit.

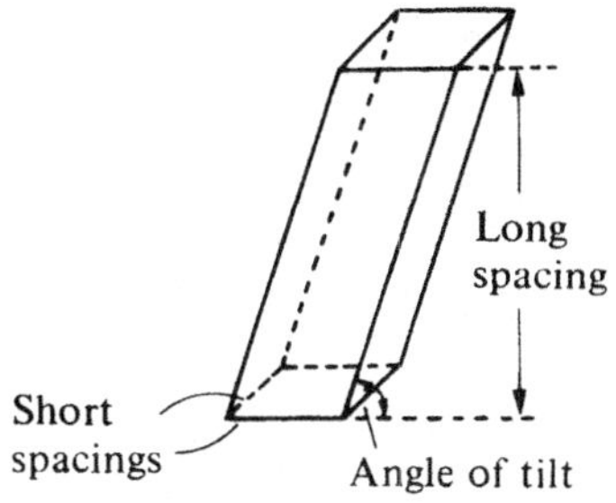

FIG. 5.1.

* Superscript numbers refer to References at end of Chapter.

2. Acids

In the following account of polymorphism in long-chain acids the descriptions odd and even refer to the number of carbon atoms in each acid molecule. The melting points of acids do not rise regularly with increasing molecular weight. Odd acids melt lower than even acids with one less carbon atom. Such values are said to show *alternation*—a phenomenon commonly displayed by the physical properties of long-chain compounds in the solid state and one that is related to the packing of molecules within the crystal. The melting points of even acids fall on a smooth curve lying above a similar and converging curve for odd acids (Fig. 5.2; some melting points are listed in Table 1.2).

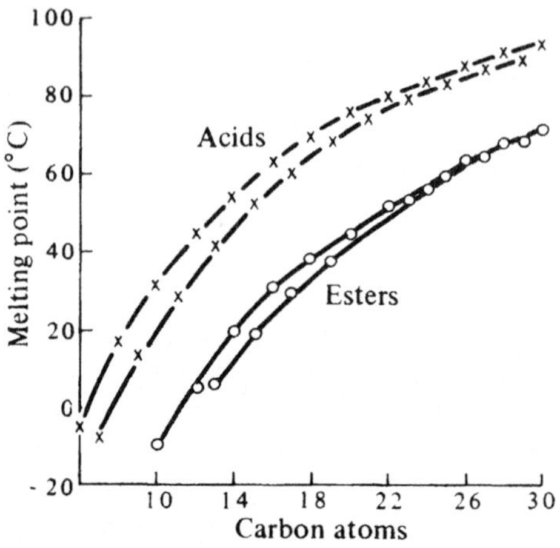

FIG. 5.2. Melting points of *n*-saturated acids and their methyl esters.

The melting points of unsaturated acids depend on the nature of the unsaturated group(s) and on their number, configuration, and relative position. Unsaturated acids of the same chain length usually have melting points in the order *cis*-olefinic < *trans*-olefinic ≃ acetylenic < saturated acid. Conjugated polyenoic acids are higher melting than their non-conjugated isomers (Fig. 5.3, Table 5.1). It follows that hydrogenation, stereomutation of *cis* double bonds, and movement of double bonds into conjugation are each accompanied by an increase in melting point. The methyl and ethyl esters of these acids, whether saturated or unsaturated, have lower melting points than the acids but show the same generalisations.

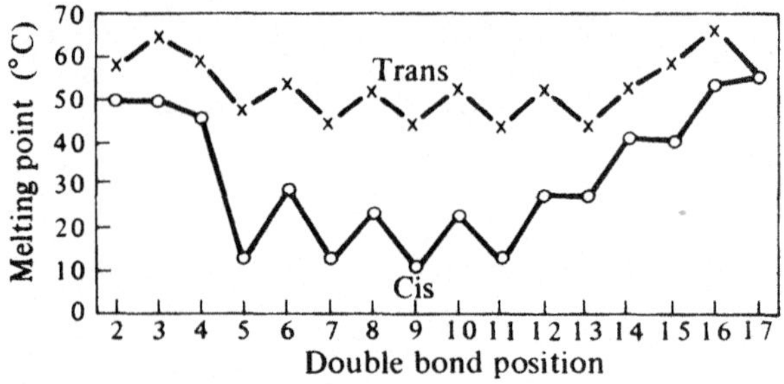

FIG. 5.3. Melting points of the octadecenoic acids.

TABLE 5.1. *The Melting Points of Some Unsaturated C_{18} Acids*

Unsaturation	M.p. (°C)	Unsaturation	M.p. (°C)
(stearic)	70	9*t*12*t*	29
9*c*	11	9*t*11*t*	54
9*t*	45	9*c*12*c*15*c*	−11
9*a*	46	9*t*12*t*15*t*	30
9*c*12*c*	−5	9*t*11*t*13*t*	71

The even saturated acids exist in at least three forms (A, B and C). The C form is obtained directly from the melt, by rapid crystallisation from solution, or by phase transformation of either the A or B forms which result by slow crystallisation from non-polar solvents. When heated, the A and B forms are transformed irreversibly to the C form, so that it is the melting point of this form which is always observed.

The odd saturated acids also exist in at least three forms (A′, B′, and C′). Cooling of the melt or crystallisation from solvent furnishes the A′ or B′ form. The higher melting C′ form is apparently obtained by phase transformation and exists just below the melting point of the acid.

The A and A′ forms show a triclinic subcellular pattern, but the remaining forms are orthorhombic.

3. Glycerides

It has been known since the classic work of Chevreul that fats exhibit a peculiarity in melting point seldom shown by other organic compounds. As far back as 1853 tristearin was reported to have three melting points of 52°, 64°, and 70°. Other monoacid triglycerides behaved similarly, though there was disagreement about how many melting points were exhibited by each compound. This phenomenon was interpreted in terms of polymorphism, i.e. the ability of a compound to exist in more than one crystalline form.

X-ray powder diffraction studies of triglycerides confirmed the existence of different crystalline forms in terms of different short and long spacings, but there was some disagreement in relating these lengths with the melting behaviour. This difficulty was resolved through the study of infrared spectra of these glycerides.

When the melt of a monoacid triglyceride is cooled quickly it solidifies in the lowest melting (α) form. When slowly heated this melts and if held just above the melting temperature it will resolidify in the β' form. In a similar way the final and stable β form can be obtained from the β' form. The β form—with the highest melting point—also results from solvent crystallisation. Some characteristics of these forms are tabulated in Tables 5.2 and 5.3.

The β form of some mixed saturated triglycerides (i.e. 16:0, 18:0, 16:0) is only obtained with difficulty, and such glycerides have a highest melting β' form. Among unsaturated glycerides only the symmetrical compounds (SUS, USU, UUU) exist in a β form: the unsymmetrical isomers (USS and UUS) have a stable β' form.

The stable β form generally crystallises in a double chain-length form (DCL or $\beta2$), but if one acyl group differs considerably from the other chains, then it will probably crystallise in a triple chain-length form (TCL or $\beta3$) since this permits more efficient packing. Such crystals display the normal β type short spacing, but have a long spacing about one and a half times as long as $\beta2$ type crystals. In the DCL form acyl groups are lined up ---1,3,2,1,3,2--- and ---2,3,1,2,3,1---

TABLE 5.2. *Characteristics of the α,β′, and β forms of Crystalline Triglycerides*

	M.p.	Short spacing(s) (nm)	Infrared absorption (cm^{-1})	Hydrocarbon chain	Subcell
α	lowest	0.4	720	perpendicular	hexagonal
β′	intermediate	0.42–0.43 and 0.37–0.40	727 719	tilted	orthorhombic
β	highest	0.46 and 0.36–0.39	717	tilted	triclinic

whereas in the TCL form the middle tier contains one type of acyl group and the upper and lower tiers both contain the other two acyl groups. Some mixed glycerides, which crystallise in TCL structures on their own, form high-melting crystals with certain other mixed glycerides (Fig. 5.4).

The glycerides 16:0, 18:1, 18:0 (POS) and 18:0, 18:1, 18:0 (SOS) which are the main components of cocoa butter crystallise in the β3 form. Any fat to be used as a cocoa butter equivalent should also show a tendency to crystallise in this form and have a similar melting behaviour. β-Tending glycerides, such as tristearin, undergo rapid expansion on crystallising. This results in a snow-like structure of low bulk density and is in contrast to the normal solid mass of β′-tending glycerides such as 16:0, 18:0, 16:0 (PSP).

The methyl groups at the top and bottom of each glyceride layer do not usually lie on a straight line, but form a boundary of a particular structure depending on the lengths of the several acyl groups. This is called the methyl terrace. The glycerides tilt with respect to their methyl end planes to give the best fit of the upper methyl terrace of one row of glycerides with the lower methyl terrace of the next row of glycerides. There are therefore several possible β2 modifications differing in the slope of the methyl terrace and in the angle of tilt.

TABLE 5.3. *The Melting Points and Long Spacings of Single Acid Triglycerides*

Acid chain length	Melting point °C			Long spacing $\times 10^{-10}$ m		
	α	β′	β	α	β′	β
8	−51	−18.0	10.0	—	—	22.7
9	−26	4.0	10.5	—	25.3	24.9
10	−10.5	17.0	32.0	30.2	27.7	26.5
11	2.5	27.0	31.0	32.7	29.8	29.6
12	15.0	34.5	46.5	35.6	32.9	31.2
13	24.5	41.5	44.5	37.8	34.2	34.0
14	33.0	46.0	58.0	41.0	37.3	35.7
15	39.0	51.5	55.0	42.9	39.2	39.2
16	45.0	56.5	66.0	45.8	42.5	40.8
17	50.0	60.5	64.0	48.5	43.8	43.5
18	54.7	64.0	73.3	50.6	47.0	45.1
19	59.0	65.5	71.0	53.1	48.1	48.2
20	62.0	69.0	78.0	55.8	50.7	49.5
21	65.0	71.0	76.0	58.5	53.2	52.7
22	68.0	74.0	82.5	61.5	56.0	54.0

[Selected from *Lipids*, 1970, **5**, 90.]

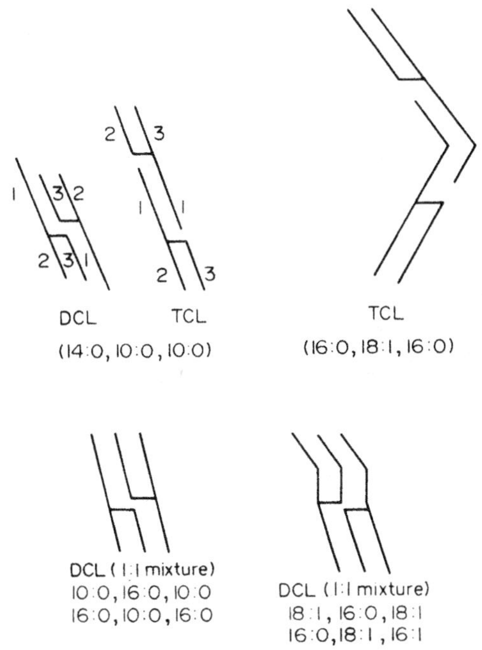

FIG. 5.4. DCL and TCL structures.

B. SPECTROSCOPIC PROPERTIES

1. Ultraviolet spectroscopy

Monoene and methylene-interrupted polyene acids absorb ultraviolet light at wavelengths too low for convenient study. Ultraviolet spectroscopy is, however, invaluable in the study of acids having a conjugated unsaturation and of reactions, such as autoxidation, producing such acids. Conjugated diene acids show a single absorption peak at 230–235 nm, whilst conjugated triene acids show three peaks at about 260, 270, and 280 nm. The actual position of absorption varies slightly with the configuration of the double bonds.

2. Infrared spectroscopy

In the solid state infrared spectra provide useful information about polymorphism, crystal structure, conformation, and chain-length, but the commonest use of infrared spectroscopy is in the recognition of *trans* unsaturation. This involves the study of liquids or solutions. One *trans* double bond produces characteristic absorption at 968 cm^{-1} and this frequency does not change for additional *trans* double bonds so long as they are not conjugated. The position of absorption for conjugated dienes, trienes, and tetraenes are slightly different from 968 cm^{-1}.

There is unfortunately no diagnostic absorption band in the infrared for *cis* unsaturation, but Raman spectra show strong absorption bands at 1656 $\pm$ 1 cm (*cis* olefins), 1670 $\pm$ 1 (*trans* olefins), and 2232 $\pm$ 1 and 2291 $\pm$ 2 (acetylenes) for the unsaturation indicated.

3. Nuclear magnetic resonance spectroscopy[3,4,5]

The ^1H nmr spectrum of a saturated ester such as methyl stearate shows four signals corresponding to the ω-methyl group (δ 0.89), the ester methyl group (δ 3.65), the α-methylene group (δ 2.21), and the remaining CH_2 groups (δ 1.26). A more powerful instrument will also show a separate signal for the β-methylene group (δ 1.58). Signals for other methylene groups close to the ester function can be distinguished with shift reagents.

An olefinic ester such as methyl oleate also shows a triplet for two olefinic hydrogen atoms (δ 5–6) and a signal for allylic hydrogen atoms (δ 1.99). Isomeric *cis*- and *trans*-alkenes differ in the coupling constant of the vinyl hydrogens (*cis* ~10 Hz, *trans* ~15 Hz) and the chemical shift of the allylic hydrogens (*cis* 1.99, *trans* 1.94). Methylene groups lying between two *cis* double bonds (as in methyl linoleate) show a signal at δ 2.72. Glycerides show signals for the $—CH_2O—$ (δ 4.2) and >CHO— (δ 5.2) hydrogen atoms (Fig. 5.5).

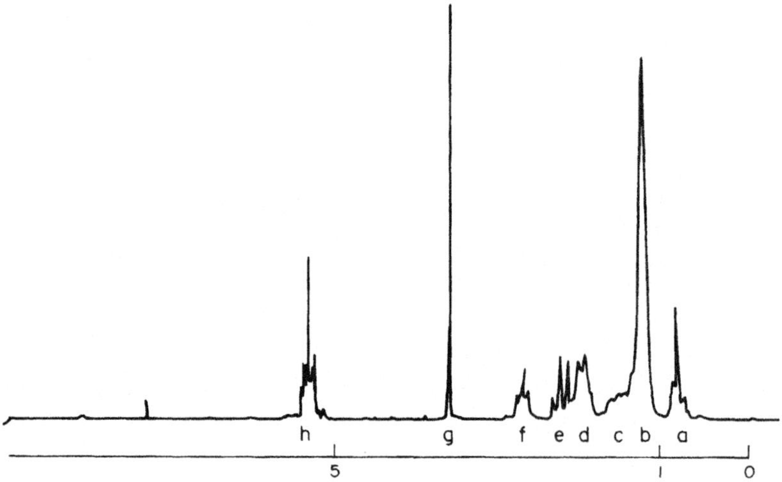

FIG. 5.5. ^1H nmr spectrum of methyl linoleate (Bruker WP80, 80 MHz spectrum)
$CH_3(CH_2)_3CH_2CH{=}CHCH_2CH{=}CHCH_2(CH_2)_4CH_2CH_2COOCH_3$
a b d h h f h h h d b c e g

The greater sensitivity of ^{13}C nmr spectra to chemical environment make these more informative than ^1H spectra despite the low abundance of ^{13}C in natural samples. A saturated methyl ester shows seven clearly separated signals along with a complex signal around 29.30 ppm in which additional signals can be distinguished with an instrument of sufficiently high resolution. Further signals are observed in unsaturated esters corresponding to the multiple bonded carbon atoms and to nearby methylene groups. It is easy to distinguish *cis* and *trans* isomers. These points are illustrated in the data given in Table 5.4 (see also Fig. 5.6).

Measurement of the solid fat content (SFC) or the solid fat index (SFI) is important in the fat industry. It is required for process control in hydrogenation, interestification, and blending. Important properties of margarine and cooking fats require close control of the SFC. A simple, quick routine method is therefore desirable.

Traditionally, this measurement has been made—though not quickly—by dilatometry. This involves a systematic investigation of the changes in volume that occur with changing temperature.

TABLE 5.4. ^{13}C *nmr Chemical Shifts for Four* C_{18} *Acids*

Acid	Carbon atom																	
	18	17	16	15	14	13	12	11	10	9	8	7	6	5	4	3	2	1
stearic	14.12	22.79	32.07	29.51				[29.82]					29.56	29.38	29.23	24.81	34.24	180.58
oleic	14.13	22.80	32.06	29.47	29.68	29.47	29.83	27.31	130.09	129.78	27.31	29.83	29.22	29.22	29.22	24.80	34.24	180.55
elaidic	14.11	22.76	32.01	29.41	29.60	29.28	29.76	32.66	130.54	130.23	32.64	29.65	29.00	29.17	29.14	24.75	34.19	180.61
linoleic	14.08	22.69	31.67	29.48	27.32	130.20	128.08	25.77	128.24	130.01	27.32	29.72	29.21	29.21	29.21	24.78	34.22	180.56

F:· D. GUNSTONE *et al.*, *Chem. Phys. Lipids*, 1976, **17,** 1; 1977, **18,** 115

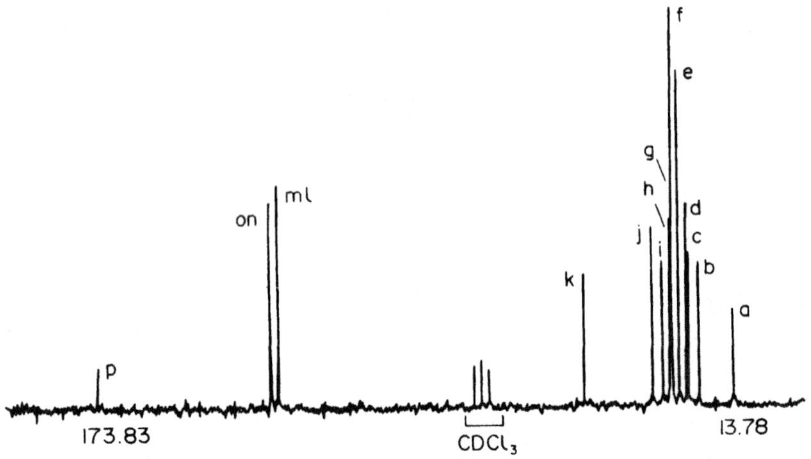

FIG. 5.6. ^{13}C nmr spectrum of methyl linoleate (Varian CFT 20, 0.5 M solution in $CDCl_3$, spectrum run at 20 MHz).
$CH_3CH_2CH_2CH_2CH_2CH{=}CHCH_2CH{=}CHCH_2CH_2(CH_2)_3CH_2CH_2COOCH_3$
a b i g e o l d m n e h f c j p k

At the present time SFC is measured by wide line nmr or by pulsed nmr. Both of these techniques distinguish hydrogen atoms in liquid and solid environments.

In wide line nmr the sample signal which measures hydrogen in a liquid environment is compared with the signal from a totally liquid sample at the same temperature. The difficulty that the sample is not completely liquid at the temperature of observation is overcome in one of two ways. Either the signal observed at a higher temperature (usually 60°) is adjusted for the change in temperature by an appropriate correction factor or the comparison is made with a liquid reference sample chosen to have about the same average unsaturation as the liquid portion of the fat under examination.

In pulsed nmr a measurement related to the total number of hydrogen atoms is followed by a second measurement 70 μs later which measures only those hydrogen atoms in a liquid environment. This determination depends on the fact that the transversal magnetisation of hydrogen in a solid environment decays much faster ($\sim 10\,\mu s$) than that of hydrogen in a liquid environment ($\sim 100\,\mu s$).

With suitable adaptation these nmr methods can be used to estimate the oil content in seeds without extraction. This is useful for plant breeders, since the seed remains viable. It may also provide a quick routine method of measuring oil composition.

4. Mass spectrometry[6,7]

The mass spectrum of a saturated ester such as methyl stearate shows a molecular ion (M^+) and peaks at $m/e = 74$ [$CH_2 = C(OH)OMe$], M-31 (M-OCH_3), $54 + 14n$ [$(CH_2)_nCOOMe$], and $15 + 14m$ [$CH_3(CH_2)_m$]. Although there is some selectivity, fission occurs between each pair of carbon atoms along the chain.

In the presence of a branched methyl group (**1**) or an oxygen-containing substituent (**2**–**5**) α-cleavage becomes more dominant and the major fragment usually indicates the nature and position of the additional functional group. α-Cleavage of ketones is accompanied by β-cleavage with McLafferty rearrangement. Fragment ions containing the methyl ester function frequently undergo further loss of 32 mass units (CH_3OH). Oxo esters are usually examined as such, hydroxy esters are converted to trimethylsilyl ethers which also have superior glc properties, and epoxy esters are converted to ethers of some kind.

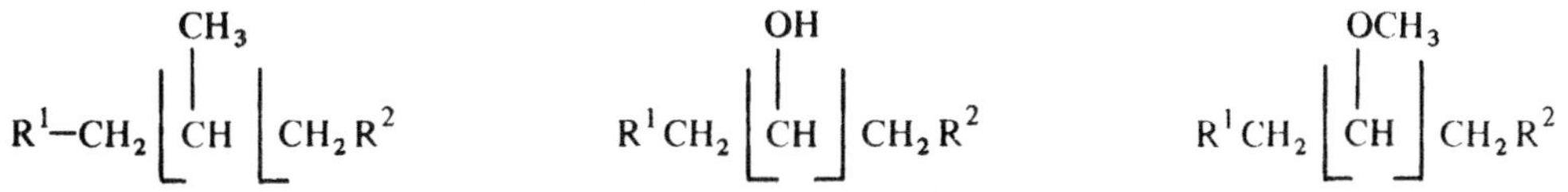

1 2 3

4 5

$R^1 = CH_3(CH_2)_m$ $R^2 = (CH_2)_nCOOCH_3$

[The symbols [and] are used to indicate the point of cleavage and the fragment which is most likely to be positively charged and so detected in the mass spectrometer.]

Although methyl oleate, linoleate and linolenate give distinctive mass spectra they are not readily distinguished from isomeric monoene, diene, and triene esters respectively because of the mobility of double bonds under electron bombardment. It is therefore necessary to "fix" the double bond position in some way. Many reactions have been recommended, but the one most commonly employed involves hydroxylation with osmium tetroxide and formation of the polytrimethylsilyl ether.

REFERENCES

1. D. CHAPMAN, *The Structure of Lipids by Spectroscopic and X-ray Techniques*, Methuen, London, 1965.
2. T. D. SIMPSON, *Fatty Acids* (ed. E. H. Pryde), American Oil Chemists' Society, Champaign, 1979, p. 157.
3. D. J. FROST and F. D. GUNSTONE, *Chem. Phys. Lipids*, 1975, **15**, 53.
4. F. D. GUNSTONE, M. R. POLLARD, C. M. SCRIMGEOUR, and H. S. VEDANAYAGAM, *Chem. Phys. Lipids*, 1977, **18**, 115.
5. D. WADDINGTON, *Fats and Oils: Chemistry and Technology* (ed. R. J. Hamilton and A. Bhati), Applied Science, London, 1980, p. 25.
6. J. A. MCCLOSKEY, *Topics in Lipid Chemistry*, 1970, **1**, 369.
7. A. ZEMAN and H. SCHARMANN, *Fette Seifen Anstrichm.*, 1972, **74**, 509; 1973, **75**, 32; 170.

CHAPTER 6

Catalytic Hydrogenation, Chemical Reduction and Biohydrogenation*

A. CATALYTIC HYDROGENATION

1. Introduction

Natural triacylglycerol mixtures are solid or liquid at room temperature. Solid fats are in great demand in most temperate climates, but liquid fats—of vegetable or fish origin—are more widely available. These can be converted to fats of higher melting range by hydrogenation (usually partial) or by blending (sometimes followed by interesterification—see Chapters 9 and 17). In addition to raising the melting range, partially hydrogenated fats are usually more stable to atmospheric oxidation and therefore less likely to develop unwanted flavours. Several million tons of oil—mainly soybean, other vegetable oils, and some fish oils— are hydrogenated annually. The fish oils contain a wide range of unsaturated acids both in respect of chain-length (C_{16}–C_{22}) and double bond number (1–6), but vegetable oils usually have oleic acid, linoleic acid, and sometimes linolenic acid, as their only unsaturated constituents.

The aim of industrial fat hydrogenation is to produce a product with the desired melting range and the required plastic properties which retains its high nutritional value. Such products result from partial hydrogenation which furnishes a more complex product than might be expected.

Hydrogenation can be effected with a range of heterogeneous and homogeneous metal catalysts including platinum, palladium, nickel, copper and cobalt, but only nickel is used extensively on an industrial scale. There may, however, be a future for copper catalysts and for some of the homogeneous catalysts which show interesting properties. Understanding of the changes which occur during the partial hydrogenation of natural unsaturated triacylglycerol mixtures has come from the careful study of the hydrogenation and deuteration of pure methyl oleate, linoleate, and linolenate.

Important features of partial hydrogenation are selectivity and isomerisation. Selectivity is concerned with the different rates of reaction of triene, diene, and monoene acyl groups, and since the term can be used in different ways it should always be defined. It may be used to describe the relative rates of hydrogenation of linolenate and linoleate, or of linoleate and oleate, or merely the production of high melting saturated and *trans*-monoene acids compared with *cis*-monoenes. During partial hydrogenation unreacted double bonds may change their position and/or their configuration. What is usually measured, however, is the total content of *trans*-acids.

* See also Chapter 14.

2. Methyl oleate

Complete hydrogenation of methyl oleate gives only methyl stearate, but incomplete reduction furnishes stearate, unchanged oleate, and a mixture of *cis*- and *trans*-octadecenoates with the double bond in several positions. This mixture has been designated iso-oleate and may contain up to twenty isomeric species (Scheme 6.1).

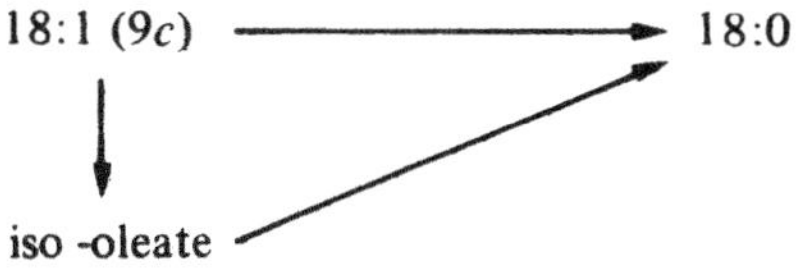

SCHEME 6.1. Hydrogenation of methyl oleate.

A comparative study of the hydrogenation of methyl oleate (18:1 9*c*) and methyl elaidate (18:1 9*t*) showed that though these two esters are hydrogenated at the same rate, double bond migration occurs more quickly in the *trans*-isomer.

3. Methyl linoleate

The isomerisation and reduction processes occurring with methyl linoleate are outlined in Scheme 6.2. The simplest sequence is reduction of diene to monoene (18:1 9*c* and 12*c*) and thence to stearate, but isomerisation reactions compete with hydrogenation yielding several dienes (conjugated and non-conjugated) and monoenes. Both double bond movement and stereomutation occur and the partially reduced product is quite complex. Its detailed composition depends on the reaction conditions (Section 5) and on the extent of reaction.

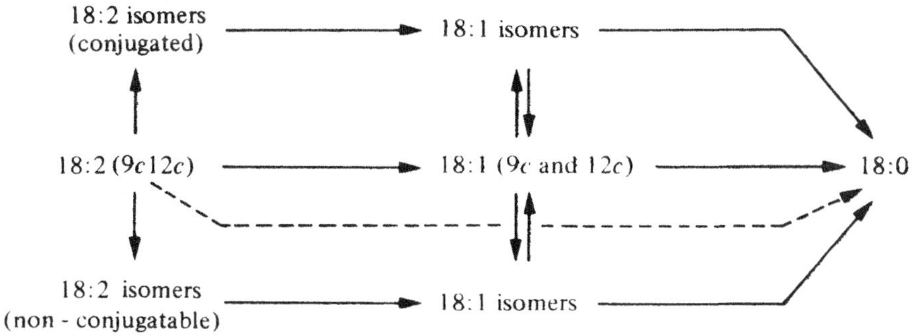

SCHEME 6.2. Hydrogenation of methyl linoleate.

The formation of conjugated dienes (mainly 9*c*11*t* and 10*t*12*c*) is particularly significant. Some catalysts, such as copper chromite, hydrogenate methylene-interrupted dienes (conjugatable) only via their conjugated isomers, and non-conjugatable dienes and monoenes are not reduced by this catalyst. Non-conjugatable dienes have more than one CH_2 group between their unsaturated centres. Reaction with nickel catalysts is not confined to this route. The quick reaction occurring via conjugated isomers is thought to account for the enhanced reactivity of methylene-interrupted polyenes compared with the monoenes and non-conjugatable polyenes. Some diene molecules are reduced directly to stearate, i.e. they undergo hydrogenation twice without desorption between each step. The full analysis of a partially reduced diene ester is quite complex and cannot be carried out in a routine manner.

4. Methyl linolenate

The reduction of methyl linolenate is even more complicated (Scheme 6.3). Direct hydrogenation furnishes, in turn, a number of methylene-interrupted (conjugatable) dienes, then monoenes, and finally stearate. Conjugation produces trienes with diene conjugation and, after more extensive double bond migration, a conjugated triene. Conjugated compounds are usually reduced more quickly than their non-conjugated isomers. The intermediate dienes are of three kinds: those with conjugated unsaturation, those with methylene-interrupted double bonds which readily yield conjugated dienes, and those with double bonds so well separated (e.g. Δ9,15 esters) that conjugation is unlikely. The first two types are readily hydrogenated, but the latter—like the monoenes— react more slowly and tend to accumulate during partial hydrogenation. All the unsaturated intermediates may undergo double bond migration and/or stereomutation instead of hydrogenation, so that many isomers other than those designated in Scheme 6.3 are actually present. Some molecules may be reduced two or even three times on a single occasion of adsorption. The relative importance of the several pathways depends on the choice of catalyst and other experimental factors.

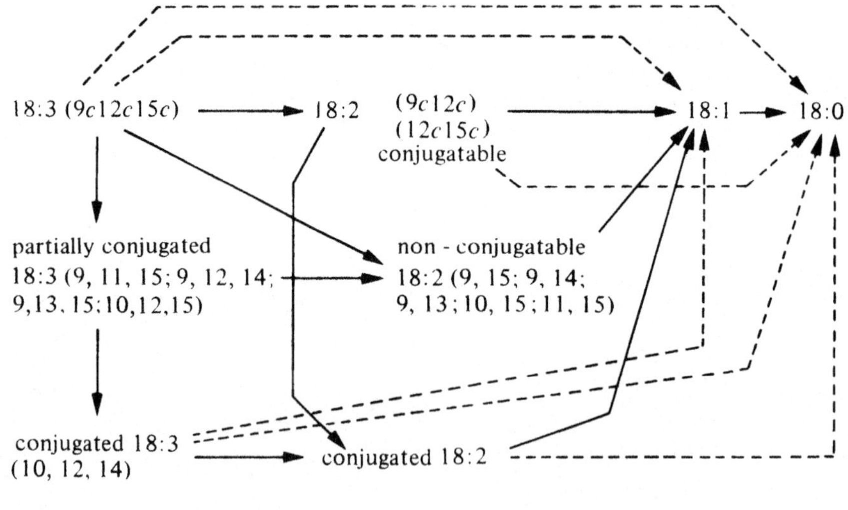

SCHEME 6.3. Hydrogenation of methyl linolenate.

Because linolenic acid (present in soybean oil) with its *n*-3 double bond furnishes undesirable flavours after oxidation, and because linoleic acid has a considerable dietary value, considerable effort has been made to produce catalysts with high linolenic/linoleic selectivity. Some success has been achieved with copper catalysts where the selectivity value can be as high as 15 compared with a value around 2 for the more commonly used nickel preparations.

5. Reaction conditions

On the commercial scale triacylglycerols, containing several different unsaturated acyl groups, are hydrogenated at pressures of 1–6 atm and temperatures of 120–200° using about 0.05% of nickel catalyst. By a suitable choice of those reaction conditions which affect the concentration of hydrogen on the catalyst, the reaction may be of low selectivity when oleic, linoleic, and linolenic acyl chains are reduced at relative rates of 1:7.5:12.5 or it may be highly selective when the relative rates are 1:50:100. Hydrogen concentration on the catalyst can be increased by decreasing the temperature, raising the pressure, increasing the rate of stirring, and decreasing the proportion of catalyst, and these factors will decrease selectivity and isomerisation. The reversal of these conditions will produce the opposite effects. Reaction conditions can therefore be selected depending on the desired nature and end-use of the hydrogenated material. There is some evidence that fatty acids in the 1- and 3- positions of a triacylglycerol are reduced more readily than similar acyl groups at position 2.

The competing processes of double bond migration and hydrogenation (or deuteration) are usually discussed in terms of the half-hydrogenated species which may suffer three different fates. They may lose hydrogen to regenerate either the original species or a position or stereoisomer or they may react with a second hydrogen atom to produce the reduced product. Alternatively π-allyl complexes might be involved. These ideas are outlined in Scheme 6.4 for the simple reaction of a monoene with deuterium which gives the saturated (dideuterio) compound along with the original alkene and a number of mono-deuterio isomers. All the monoene products can re-enter a similar reaction cycle, leading eventually to extensive double bond movement and stereomutation (and deuterium incorporation). Up to 30 atoms of deuterium may be introduced into methyl stearate ($C_{17}H_{35}COOMe$) during deuteration of oleate. This is an indication of the extent to which hydrogen exchange occurs.

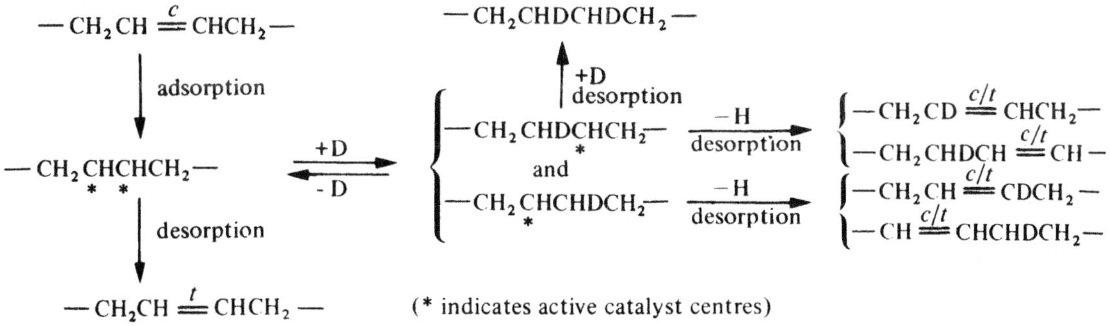

SCHEME 6.4. Monoene deuteration via half-deuterated species.

Research on the use of homogeneous catalysts for hydrogenation has produced some interesting results, but such catalysts are not yet being used on a commercial scale. The catalysts examined include cyanocobalt complexes, transition metal acetylacetonoates (Ni, Co, Cu, Fe), metal triphenylphosphine complexes (Pt, Pd, Ni), and metal carbonyls (Fe, Co, Mn, Cr, Co, W).

Iron pentacarbonyl, for example, catalyses most of the reactions observed with heterogeneous catalysts. It is, however, selective for polyenes and does not reduce monoenes. Chromium carbonyl complexes reduce polyenes to *cis*-monoenes only. Linoleate, for example, gives the Δ9, 10, 11 and 12 *cis*-monoenes.

B. CHEMICAL REDUCTION

The conversion of an unsaturated acid or ester to its fully hydrogenated derivative—sometimes a significant step in the identification of acids of unknown structure—is readily achieved by catalytic hydrogenation using a platinum, palladium, or nickel catalyst. The partial and stereospecific reduction of alkynes to *cis*-alkenes—important in the synthesis of olefinic acids—is most commonly effected with a palladium catalyst deposited on calcium carbonate containing some lead (Lindlar's catalyst) in the presence of quinoline to enhance selectivity.

Non-catalytic reduction of alkenes is conveniently carried out with hydrazine (N_2H_4). Oxygen is necessary and the active reagent is probably di-imine (N_2H_2). The reaction is stereospecific (*cis*-addition) and occurs without double bond migration or stereomutation. Partial reduction of a polyene therefore gives a simpler product than the catalytic processes. A typical experiment with linolenic acid is summarised in Scheme 6.5. The stereospecific *cis* addition provides the best route to *vic*-dideuterio acids of known stereochemistry.

$$\underset{cis\ \text{or}\ trans}{-\mathrm{CH{=}CH}-} \xrightarrow{N_2D_4} \underset{erythro\ \text{or}\ threo}{-\mathrm{CHDCHD}-}$$

18:3	→ 18:2	→ 18:1	→ 18:0 (9)
9*c*12*c*15*c* (26%)	9*c*12*c* (15%)	9*c* (9%)	(5%)
	12*c*15*c* (15%)	12*c* (9%)	
	9*c*15*c* (13%).	15*c* (8%)	

SCHEME 6.5. Partial reduction of methyl linolenate with hydrazine.

C. BIOHYDROGENATION

Ruminant animals such as sheep and cattle differ from other animals in that food digestion starts in the rumen where microorganisms hydrolyse dietary lipids and hydrogenate linoleic and linolenic acid to stearic acid and to unsaturated acids with double bonds in unusual positions and mainly with *trans*-configuration. The rumen bacterium, *Butyrivibrio fibrosolvens*, for example, contains an isomerase and a reductase which together convert linoleic acid to vaccenic acid thus:

$$18{:}2\ (9c12c) \xrightarrow{\text{isomerase}} 18{:}2\ (9c11t) \xrightarrow{\text{reductase}} 18{:}1\ (11t)$$

Another rumen bacterium which reduces oleic and linoleic acid to stearic acid converts linolenic acid to the 18:1 (15*c*) acid.

Rumen hydrogenation can be circumvented by a special feeding regimen in which unsaturated acids are protected during their passage through the rumen. Both the meat and the milk then contain more linoleic acid than usual, and the milk can be used to produce a wide range of polyunsaturated dairy products.

GENERAL REFERENCES

DUTTON, H. J., *Progress in the Chemistry of Fats and Other Lipids*, 1968, **9**, 349.
FRANKEL, E. N., and H. J. DUTTON, *Topics in Lipid Chemistry*, 1970, **1**, 161.
MOUNTS, T. L., *Fatty Acids* (ed. E. H. Pryde), American Oil Chemists' Soc., Champaign, 1979, p. 391.

CHAPTER 7

Oxidation*

A. OXIDATION BY OXYGEN

1. Introduction

Oxidation of olefinic compounds by atmospheric oxygen is important in the development of rancidity, in the production of desirable and undesirable flavours, in the polymersiation of highly unsaturated (drying) oils, and in the production of compounds of significant physiological activity. These changes, which may or may not require enzymes, result from complex reactions in complicated substances often under undefined conditions and understanding of the processes involved has developed from the study of simpler substrates such as methyl oleate, linoleate, or linolenate.

Reaction between olefin and oxygen probably requires the activation of the alkene or of the oxygen, and the two processes follow different pathways to slightly different products. The first isolable oxidation products are unsaturated hydroperoxides which undergo further reaction to produce compounds of lower molecular weight after chain fission, compounds of similar molecular weight, and compounds of higher molecular weight after dimerisation or polymerisation.

2. Autoxidation[1,2]†

The major non-enzymic oxidation process is a radical chain reaction involving initiation, propagation, and termination steps (Scheme 7.1).

initiation	production of R· or RO_2· radicals
propagation	$R\cdot + O_2 \longrightarrow RO_2\cdot$
	$RO_2\cdot + RH \longrightarrow RO_2H + R\cdot$
termination	interaction of radicals to produce non-initating and non-propagating products
	(RH represents alkene substrate)

SCHEME 7.1. Autoxidation of the alkene RH, where H represents an allylic hydrogen.

* See also Chapter 19.

† Superscript numbers refer to References at end of Chapter.

The nature of the initiation reaction is still uncertain, though it is known that hydroperoxides once formed furnish additional initiating radicals. The reaction of alkenes with singlet oxygen (Section 3) to produce hydroperoxides may play a key role in the initiation of autoxidation. The propagation sequence involves the production of a radical R· from the alkene RH and its subsequent reaction with oxygen. The radical results from the alkene by reaction at the allylic position and is resonance-stabilised. This affects the structure of the final product. The termination reactions have not been extensively studied.

Autoxidation is facilitated by pro-oxidants and inhibited by antioxidants. Pro-oxidants, such as metals or other radical initiators, operate by promoting the initiation step in the chain reaction or they may inhibit the activity of antioxidants. Antioxidants are frequently added to fats and to fat-containing foodstuffs to prolong shelf-life. These are often phenolic compounds, but only approved substances may be added to materials which are to be eaten. Such compounds interfere with the propagation sequence by converting propagating radicals into non-propagating species. Their effectiveness is often increased by compounds such as citric acid, ascorbic acid, or phosphoric acid (called synergists), all of which inhibit the initiation step by removal of metallic impurities which otherwise act as pro-oxidants. Photo-oxidation is inhibited by singlet oxygen quenchers such as carotene. (For further discussion see Chapter 19.)

Autoxidation of pure methyl oleate (in particular, oleate which is free of linoleate) is a slow reaction, occurring only after a long induction period. This can be shortened by addition of a radical source, by irradiation, or by raising the temperature. Samples of oleate containing linoleate also have shorter induction periods because the more readily formed products of linoleate oxidation can initiate oleate oxidation. The hydroperoxides produced from methyl oleate are a mixture of the *cis* and *trans* isomers of 8-hydroperoxy Δ9-, 9-hydroperoxy Δ10-, 10-hydroperoxy Δ8-, and 11-hydroperoxy Δ9-octadecenoates. The formation of these eight products is explained in terms of the propagation sequence occurring via two resonance-stabilised allyl radicals (Scheme 7.2). Recent studies have shown that the 8- and 11-hydroperoxides are each formed in ~27% yield with *cis* and *trans* isomers at about equal level and the 9- and 10-hydroperoxides are each formed in 23% yield and are almost entirely *trans* isomers. These figures require some refinement of the classical mechanism outlined in Scheme 7.2.

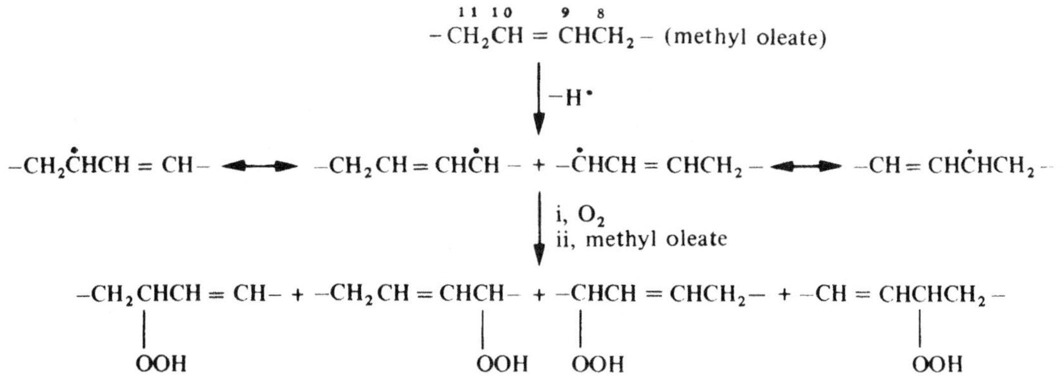

SCHEME 7.2. Autoxidation of methyl oleate.

Methyl linoleate reacts 10–40 times quicker than oleate because of the enhanced reactivity of the C-11 methylene group lying between the double bonds. The reaction product is a mixture of 9- and 13-hydroperoxyoctadecadienoates and there is no firm evidence for the formation of the 11-

hydroperoxy isomer. The *cis trans* conjugated dienes produced at 0° isomerise readily to *trans trans* dienes at room temperature and above (Scheme 7.3). Recent studies have shown that the 9- and 13-hydroperoxides rearrange, possibly by the mechanism shown in Scheme 7.4. This is a reversal of the procedure by which the hydroperoxides are formed.

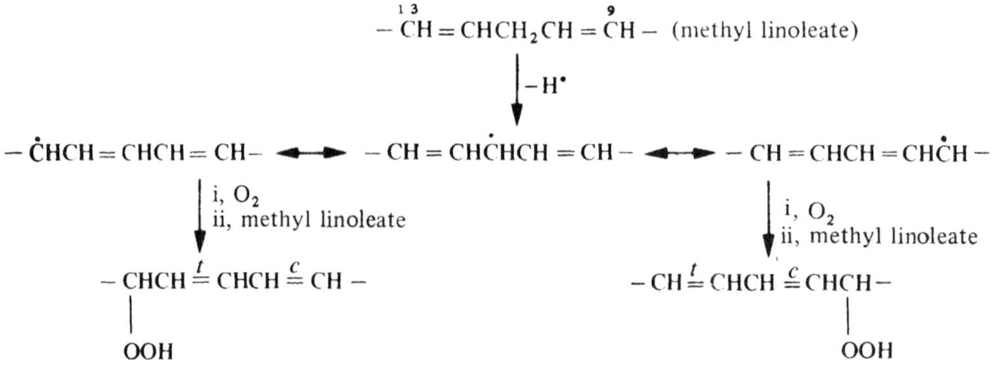

SCHEME 7.3. Autoxidation of methyl linoleate.

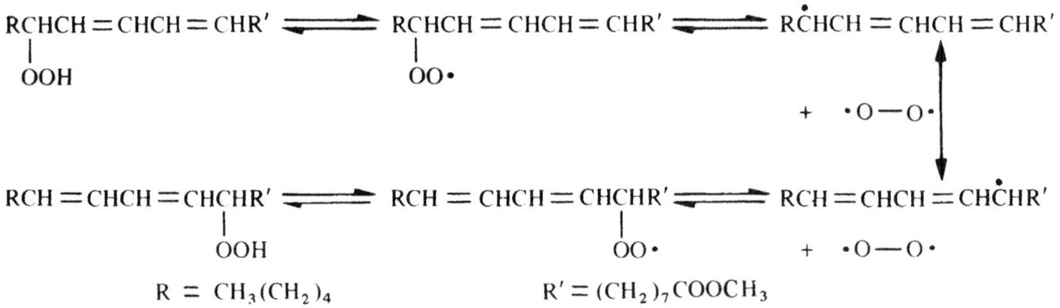

SCHEME 7.4. Rearrangement of methyl 9- and 13-hydroperoxyoctadecadienoates.

Methyl linolenate reacts 2–4 times quicker than methyl linoleate. A typical ratio for the three isomers is 1:27:77. Like linoleate, it undergoes oxidation of its 9,12 and 12,15 diene systems to yield four major hydroperoxides with three unsaturated centres, including a conjugated diene (Scheme 7.5). The 9- and 16-hydroperoxides are formed in similar amounts and make up 75–82% of these simple hydroperoxides; the 12- and 13-hydroperoxides, also formed in similar amount, make up the remaining 18–25%. The intermediate peroxy radicals leading to the inner hydroperoxides can also react with an adjacent double bond to produce cyclic peroxides (epidioxides) such as **1–3**.

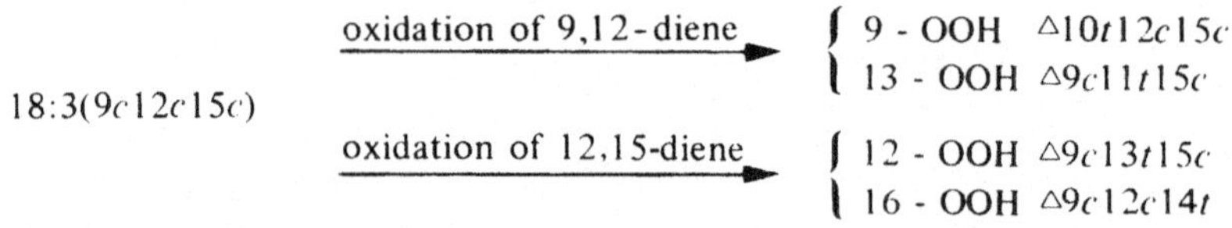

SCHEME 7.5. Autoxidation of methyl linolenate.

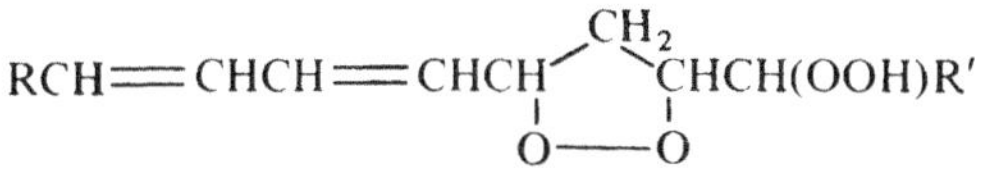

1

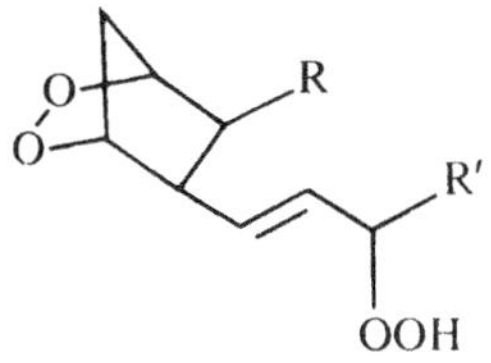

2

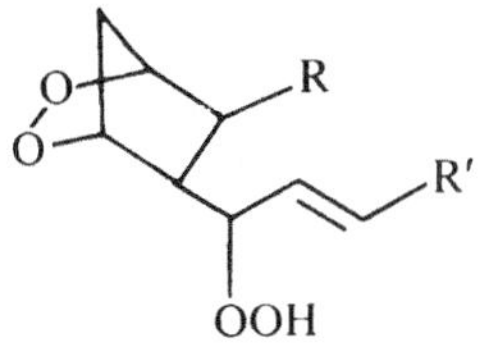

3

$R = CH_3CH_2$, $R' = MeO_2C\,(CH_2)_7$ or $R = MeO_2C(CH_2)_7$, $R' = CH_3CH_2$

3. Reaction with singlet oxygen

Photo-oxidation involves reaction of an alkene with oxygen in the presence of light and a suitable sensitiser. Sensitisers such as riboflavin activate the alkene and produce the same products as autoxidation, but other sensitisers such as erythrosine or methylene blue convert oxygen to its more reactive singlet state. This reacts with alkenes in a non-radical concerted process (the ene reaction): oxygen becomes attached to one of the unsaturated carbon atoms and the reaction is accompanied by double bond migration.

Photo-oxidation is much quicker than autoxidation and the difference in reactivity between oleate, linoleate, and linolenate (1:1.3:2.3) is close to the number of double bonds in these esters. The hydroperoxides produced in this way differ from those resulting from autoxidation.

methyl oleate	9-OOH Δ10		
	10-OOH Δ8		
methyl linoleate	9-OOH Δ10,12	12-OOH Δ9,13	
	10-OOH Δ8,12	13-OOH Δ9,11	
methyl linolenate	9-OOH Δ10,12,15	12-OOH Δ9,13,15	15-OOH Δ9,12,16
	10-OOH Δ8,12,15	13-OOH Δ9,11,15	16-OOH Δ9,12,14

It has been suggested that autoxidation of natural oils may be initiated by photo-oxidation due to pigments remaining in the oil even after processing.

4. Enzymic oxidation[3,4]

Lipoxygenase—widely distributed throughout the plant kingdom but also existing in animals—promotes reaction between oxygen and some unsaturated acids. The natural substrate appears to be linoleic acid, but other acids are also oxidised.

With lipoxygenase-1 from soybean, linoleic acid is oxidised to the 13*S*-hydroperoxy Δ9*c*11*t* acid. Other enzyme preparations, such as those from tomato and potato, give the 9*R*-hydroperoxy Δ10*t*12*c* acid, and sometimes mixtures of the two optically active hydroperoxides are obtained. In contrast, autoxidation produces racemic products. Other *n*-6 acids furnish the appropriate *n*-6 or *n*-10 hydroperoxy acids. Among the complete series of isomeric 18:2 acids (2*c*5*c* to 14*c*17*e*), only the 9*c*12*c* and, to a lesser extent, the 13*c*16*c* dienes are oxidised under the influence of lipoxygenase. The 13*c*16*c* gives the 17*S*-hydroperoxide along with a small amount of the 13 isomer.

Lipoxygenase contains one atom of iron in its molecule and exists in three states described as the colourless (native) enzyme, the yellow enzyme, and the purple enzyme. It promotes both aerobic and anaerobic reactions. The aerobic reaction starts with the removal of hydrogen from C-11. Removal of the 11*S* hydrogen is followed by oxygen insertion at C-13; removal of the 11*R* hydrogen is followed by oxygen insertion at C-9. The reaction occurs by the following oxidation sequence (see also Scheme 7.6).

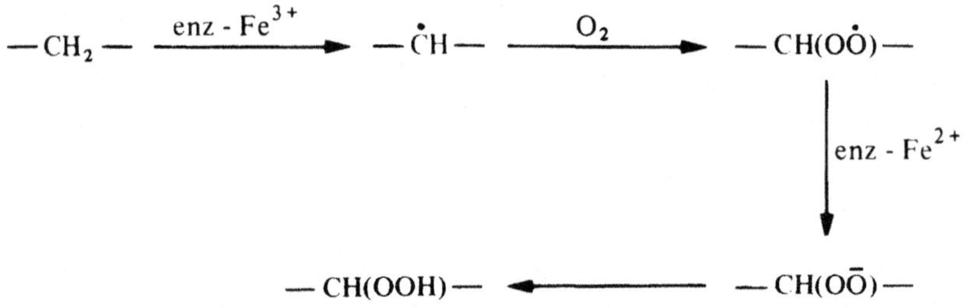

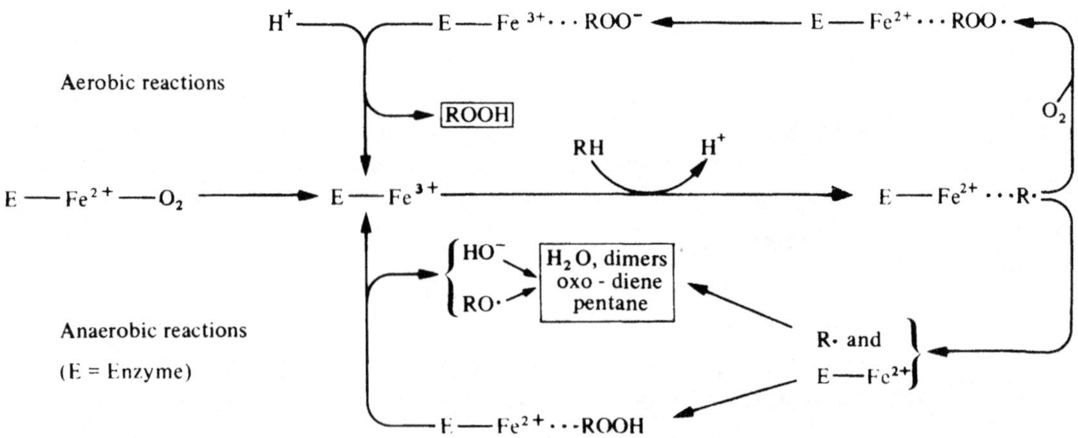

SCHEME 7.6. Lipoxygenase-catalysed oxidation of linoleic acid under aerobic and anaerobic reaction conditions.

When there is a deficiency of oxygen the reaction follows a different course. Unreacted linoleic acid interacts with the 13-hydroperoxide to give a mixture of products including pentane, the C_{13} aldehyde (**4**), the C_{18} oxo diene (**5**), and dimeric products.

$$\mathrm{OHCCH \overset{t}{=} CHCH \overset{c/t}{=} CH(CH_2)_7COOH} \qquad \mathrm{CH_3(CH_2)_4COCH \overset{t}{=} CHCH \overset{t}{=} CH(CH_2)_7COOH}$$

4 **5**

In addition to the expected hydroperoxides, linolenic acid, in the presence of soy flour, gives the cyclic peroxyhydroperoxides (**6**) identical with those (**1**) observed in the autoxidation of methyl linolenate.

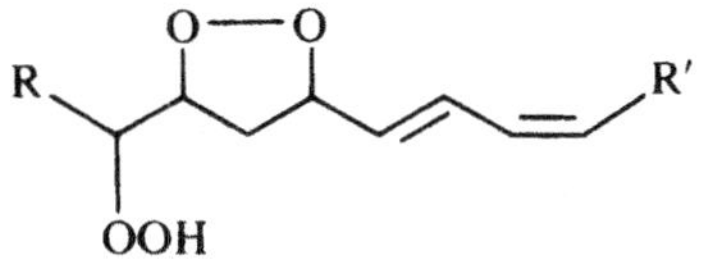

$\mathrm{R = CH_3CH_2 \quad R' = (CH_2)_7COOH}$

or $\mathrm{R = (CH_2)_7COOH \quad R' = CH_3CH_2}$

6

In animal systems of enzymic oxidation the most common substrate is arachidonic acid (20:4). One group of enzymes produces the prostaglandins and thromboxanes (Chapter 4) whilst animal-derived lipoxygenases produce the leukotrienes, another family of physiologically active compounds.

5. Reactions of hydroperoxides

Short-chain products resulting from fission of the C_{18} chain include hydrocarbons (such as ethane and ethene from linolenic and other *n*-3 polyene acids and pentane from linoleic and other *n*-6 polyene acids), aldehydes, ketones, esters, lactones, alcohols, and ethers, all of which may be saturated or unsaturated. Typical aldehydes include alkanals, alkenals (Δ2, Δ3, Δ4), alkadienals (Δ2,4; Δ2,5; Δ2,6), and alkatrienals (Δ2,4,7), all of which may affect flavour in a desirable or undesirable way. In the complex question of flavour, human response varies with the concentration of a particular component and on the presence of other flavour-producing compounds. The threshold of organoleptic observation of many of these compounds is measured in parts per million or per billion, so that exceedingly small amounts may have a significant effect on flavour. The formation of many of these compounds is easily rationalised in terms of the structure of the original hydroperoxides and the major pathways of breakdown. But some are not, and it is uncertain whether these arise from unexpected substrates or from unrecognised breakdown pathways. Hydroperoxides of unexpected structure could result from acids of conventional structure by an unusual oxidation route or by the standard oxidation of a minor fatty acid not yet recognised in the substrate. Fish oils and vegetable oils—even before the structural complications produced during partial hydrogenation (Chapter 6)—may contain traces of unidentified acids with double bonds in unexpected positions.

The decomposition of hydroperoxides to short-chain aldehydes occurs by homolytic or heterolytic processes leading to slightly different products (Scheme 7.7). Homolytic cleavage of the 13- and 9-hydroperoxides from linoleic ester would explain the presence of the 6:0 and 10:2 (2*t*4*c*)

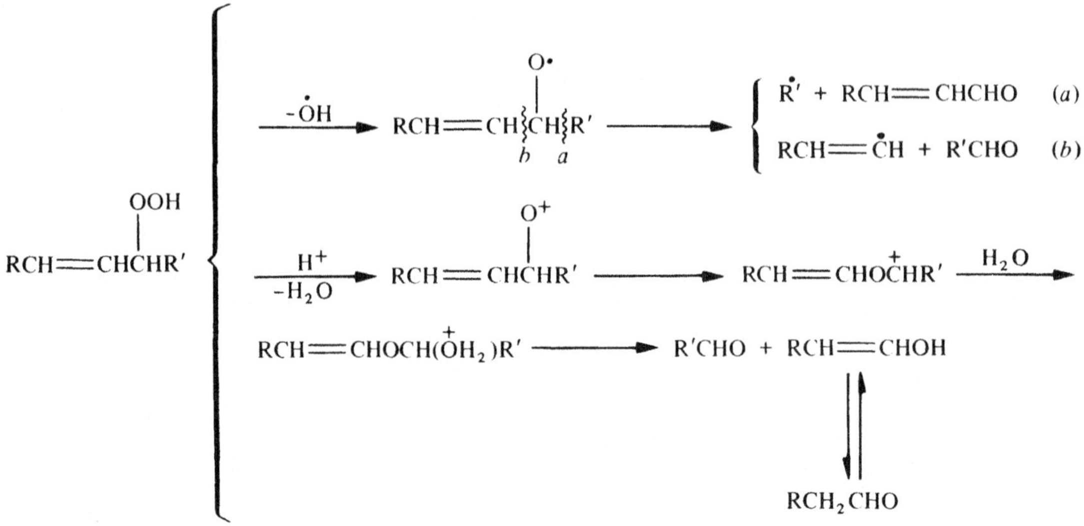

SCHEME 7.7. The formation of short-chain aldehydes from allylic hydroperoxides.

aldehydes. The latter conveys a "deep fried" flavour at a threshold concentration one in 10^{10}. The four hydroperoxides from oleic acid/ester (Scheme 7.2) give the 8:0, 9:0, 10:1(2*t*) and 11:1(2*t*) aldehydes. The situation is more complicated with linolenic acid and with partially hydrogenated acids.

The hydroperoxides are also changed to compounds of similar chain-length under both enzymic and non-enzymic conditions. Among the compounds arising from the 9- and 13-linoleate hydroperoxides are hydroxy dienes (**8**), oxo dienes (**9**), α- and γ-ketols and their derivatives (**10**), epoxides (**11**), and the vinyl ether (**12**). Part of the product in **10** results from solvents or other reactants also present.

RCH(OH)CH=CHCH=CHR′

8

RCOCH=CHCH=CHR′

9

RCH(OZ)COCH$_2$CH=CHR′

RCH$_2$COCH=CHCH(OZ)R′

Z = H, OR″, or OCOR″

10

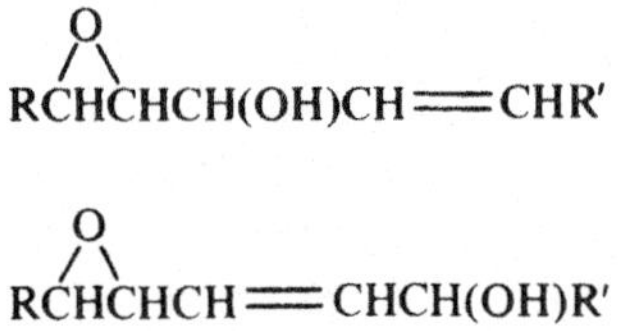

11

RCH=CHCH=CHOCH=CHR′

12

8 - 11 R = $CH_3(CH_2)_4$, R′ = $(CH_2)_7COOH$ or

R = $(CH_2)_7COOH$, R′ = $CH_3(CH_2)_4$

12 R = $CH_3(CH_2)_4$, R′ = $(CH_2)_6COOH$

The 6:1(Δ2), 9:1(Δ2), and 9:2(Δ2,6) unsaturated aldehydes derived from linoleic and linolenic acid are important components in the characteristic flavour of cucumber. When the cells are broken during cutting or eating, enzymes come into contact with unsaturated acids and the unsaturated aldehydes are quickly liberated. Three enzymic processes are involved in this change: lipolysis, hydroperoxidation, and hydroperoxide cleavage.

B. EPOXIDATION[5,6]

The oxidation of unsaturated acids and esters to produce epoxides is an important reaction. Epoxidised oils, produced at the rate of $\sim 10^8$ lb/yr, are used mainly as stabiliser–plasticiser for PVC. Natural epoxy acids are also known (Chapter 1) and are usually optically active.

Epoxidation is conveniently effected by a wide range of peracids. Unless carefully buffered, the lower aliphatic members (RCO_3H, R = H, CH_3, CF_3) give acylated diols through acid-catalysed ring-opening of the epoxide, but perlauric, monopersuccinic, monoperphthalic, and perbenzoic smoothly effect epoxidation. The commercially available *m*-chloroperbenzoic acid is widely employed on a laboratory scale.

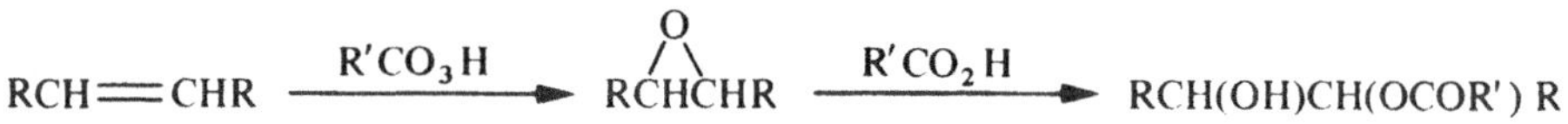

The reaction is a *cis*-addition so that *cis*- and *trans*-epoxides result from *cis*- and *trans*-alkenes respectively. Oleic acid and its *trans*-isomer (elaidic acid) furnish the *cis*- (m.p. 59.5°) and *trans*- (m.p. 55.5°) epoxystearic acids respectively. These molecules contain two chiral centres, but the products are racemic.

Partial epoxidation of methyl linoleate gives the two monoepoxyoctadecenoates and complete reaction furnishes *cis*-9,10; *cis*-12,13-diepoxystearate, which exists in two stereoisomeric forms melting at 78° and 41° (acids) and at 32°, and oil (methyl esters).

Epoxides are reactive compounds, especially in acidic solution. They can be converted into diols or their derivatives, can furnish the alkenes from which they are prepared, and readily undergo cleavage.

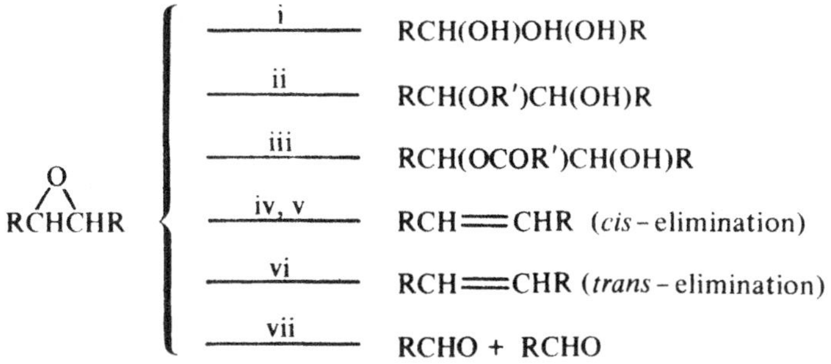

i, H_3O^+; ii, ROH,H^+; iii, R'COOH,H^+; iv, NaI, NaOAc, MeCOOH or EtCOOH; v, $SnCl_2$, $POCl_3$, pyridine; vi, $LiPPh_2$, MeI; vii, HIO_4.

SCHEME 7.8. Reaction of epoxides.

C. HYDROXYLATION

The conversion of an alkene to its diol (hydroxylation) can be carried out by several reagents. Those that effect stereospecific addition furnish *threo*- or *erythro*-diols depending on the nature of the addition process (*cis* or *trans*) and the configuration of the alkene (*cis* or *trans*).

—CH(OH)CH(OH)— ← (*trans*-hydroxylation) —CH═CH— (*cis*-hydroxylation) → —CH(OH)CH(OH)—

threo - diol	*cis* - alkene	*erythro* - diol
erythro - diol	*trans* - alkene	*threo* - diol

Potassium permanganate and osmium tetroxide effect *cis*-hydroxylation by reactions which occur through similar cyclic intermediates (Scheme 7.9). Reaction with permanganate furnishes the diol directly, but the intermediate osmate ester has to be decomposed by sodium sulphite, methanal (formaldehyde), hydrogen sulphide, or excess mannitol. This reaction occurs in high yield and is simple to perform, but suffers from the disadvantage that osmium tetroxide is expensive and toxic. Modified procedures employ osmium tetroxide as a catalyst in the presence of an oxidising agent to continually regenerate the tetroxide. Metal chlorates, hydrogen peroxide-*t*-butanol, *t*-butylhydroperoxide, and N-methylmorpholine N-oxide have been successfully used for this purpose.

—CH(OH)CH(OH)— ← (*trans* hydroxylation) —CH═CH— (*cis* hydroxylation) → —CH(OH)CH(OH)—

threo-diol	*cis* -alkene	*erythro* -diol
erythro-diol	*trans* - alkene	*threo* -diol

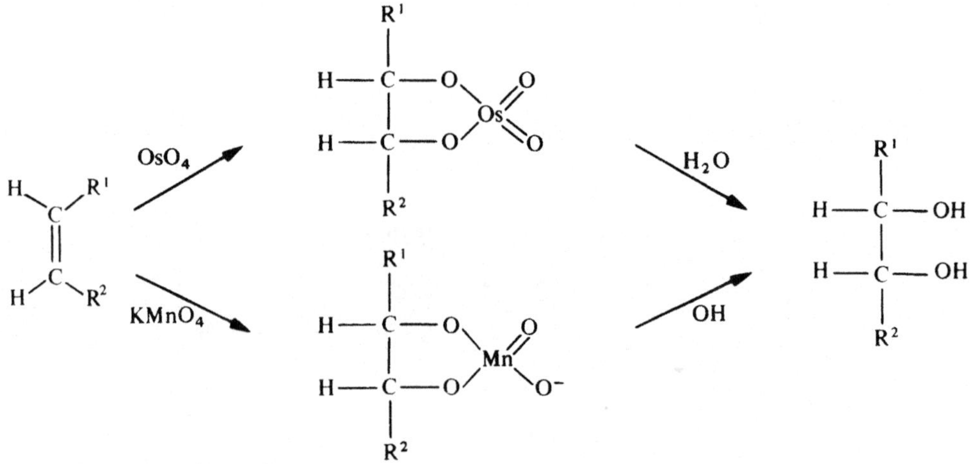

SCHEME 7.9. *cis*-Hydroxylation of a *cis*-alkene.

Halogens and metal salts are used in different ways for *cis*- and *trans*-hydroxylation. *cis*-Hydroxylation (Woodward procedure) occurs through reaction of an alkene with iodine and silver acetate in *wet* acetic anhydride, though better results are claimed with cupric or potassium salts in place of silver acetate. The Prévost reaction, using iodine and silver benzoate under *anhydrous* conditions, leads to *trans*-hydroxylation.

$$\text{—CH=CH—} \xrightarrow[\textit{trans}\text{ - addition}]{\text{RCOOAg, I}_2} \text{—CHICH(OCOR)—} \xrightarrow[\text{Prēvost}]{\text{Woodward}}$$

$$\text{—CH(OCOR)CH(OH)—} \xrightarrow{\text{hydrolysis}} \text{—CH(OH)CH(OH)—}$$

trans-Hydroxylation is more usually carried out with peracids. These furnish epoxides (Section B) which can be subjected to acid-catalysed hydration, a reaction which occurs with inversion. This is conveniently achieved in a one-pot reaction with the lower aliphatic peracids (peracetic, performic) prepared *in situ* by mixing the carboxylic acid with hydrogen peroxide and an acidic catalyst. Monoepoxides are smoothly hydrated to *vic*-diols but bis-epoxides from methylene-interrupted dienes furnish 1,4- and 1,5-epoxides in addition to the desired tetrahydroxy compounds.

$$\text{—CH=CH—} \xrightarrow[\textit{cis}\text{ - addition}]{\text{RCO}_3\text{H}} \text{—}\overset{\text{O}}{\text{CHCH}}\text{—} \xrightarrow{\text{RCO}_2\text{H}} \text{—CH(OH)CH(OCOR)—}$$

$$\xrightarrow{\text{hydrolysis}} \text{—CH(OH)CH(OH)—}$$

Simple dihydroxy esters contain two chiral centres, but the products of hydroxylation are usually racemic. Oleic acid gives *erythro*- (m.p. 132°) and *threo*- (m.p. 95°) dihydroxystearic acid. 9,10,12,13-Tetrahydroxystearic acid—from octadeca-9,12-dienoic acid—can exist in eight racemic forms. Two (m.p. 174 and 164°) result from *cis*-hydroxylation of linoleic acid and another pair (m.p. 148 and 126°) from *trans*-hydroxylation of the same acid. The same four acids can be obtained from linelaidic acid (9*t*12*t* diene). The remaining isomers could be obtained from the 9*c*12*t* or 9*t*12*c* diene acids. Ricinoleic acid (12*R*-hydroxyoleic acid) with its C-12 chiral centre yields four enantiomeric 9,12,13-trihydroxystearic acids which have been separated and characterised.

The *vic*-diols are cleaved by periodic acid to aldehydes and by permanganate to acids. Alkenes can be regenerated stereospecifically as shown in the following equations:

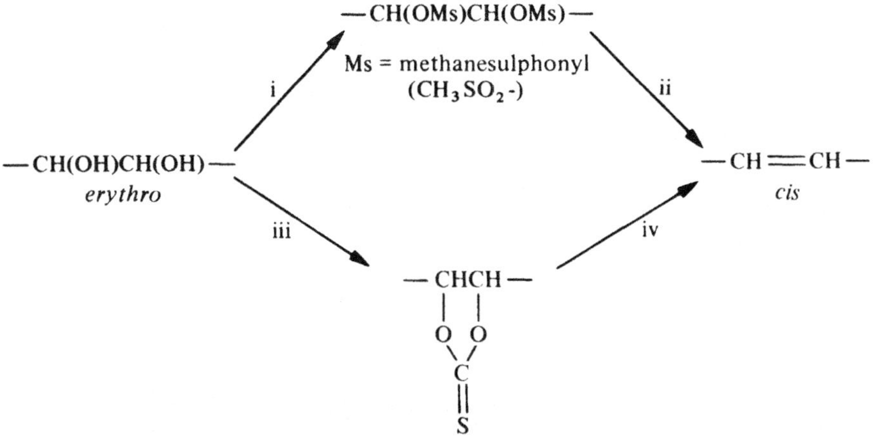

i, MsCl; ii, NaI, Zn, DMF; iii, thiocarbonyldi-imidazole, iv, $P(OMe)_3$

D. OXIDATIVE FISSION

Unsaturated acids undergo cleavage at their unsaturated centres in a number of oxidation reactions. For structural identification the methods employed should give clear results with the minimum quantity of material and provide an unambiguous solution. When complete oxidation does not lead to an unequivocal structure then partial oxidation or reduction may be employed. Such problems are now usually solved by spectroscopic procedures (Chapter 5). Only two oxidation procedures are now commonly employed: von Rudloff oxidation and ozonolysis. Oxidative cleavage is also of preparative value in the laboratory and is used on an industrial scale for the preparation of dibasic acids and, to a lesser extent, other difunctional compounds.

Vigorous oxidation with potassium permanganate results in over-oxidation, but this difficulty is overcome in the von Rudloff procedure by using as oxidant a 1:39 (molar) mixture of potassium permanganate and sodium metaperiodate. Under these conditions the concentration of permanganate remains low throughout the reaction. The reaction proceeds in aqueous solution for acids and in aqueous *t*-butanol for esters at room temperature during 6–24 hours through the stages of diol or ketol formation, cleavage, and oxidation as shown in the following sequence. The excess of periodate continually regenerates the permanganate.

$$R^1CH{=}CHR^2 \xrightarrow{KMnO_4} \left\{ \begin{array}{l} R^1CH(OH)CH(OH)R^2 \\ R^1COCH(OH)R^2 \\ R^1CH(OH)COR^2 \end{array} \right\} \xrightarrow{NaIO_4} \begin{array}{l} R^1CHO + R^2CHO \\ R^1COOH + R^2CHO \\ R^1CHO + R^2COOH \end{array}$$

$$\xrightarrow{KMnO_4} R^1COOH + R^2COOH$$

Ozonolysis is now more commonly employed both for analytical and preparative purposes. Reaction occurs via the molozonide and the Criegee zwitterion which reacts further, depending on the reaction conditions, to give an ozonide, an alkoxyhydroperoxide, or an acyloxyhydroperoxide. Each of these, however, give the same fission products which may be alcohols, aldehydes, acids, or amines.

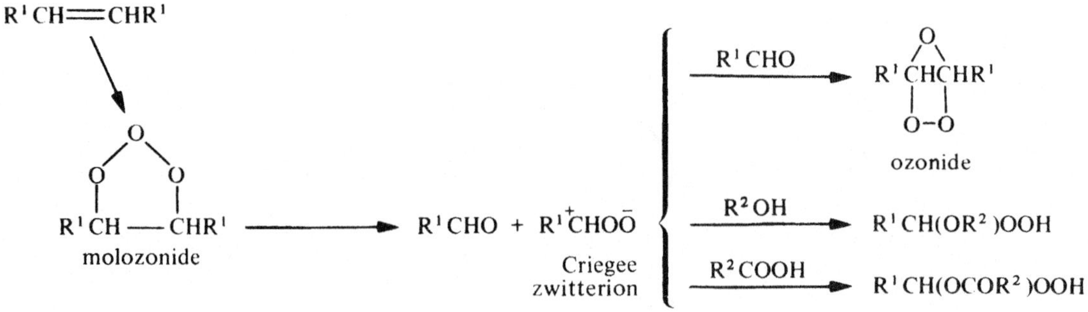

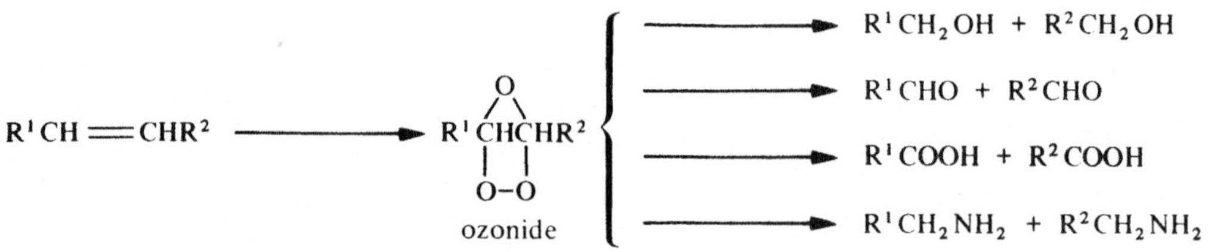

Alcohols result from reduction of the ozonide (or other product) with metal hydrides (lithium aluminium hydride, sodium borohydride) or by catalytic hydrogenation (nickel, platinum). Aldehydes are produced under milder reducing conditions with zinc-acid, triphenylphosphine, dimethyl sulphide, or hydrogen and Lindlar's catalyst. Acids are formed with a wide range of oxidising agents including peracids and silver oxide. Amines result when ozonides are reduced over Raney nickel in the presence of ammonia or can be obtained from aldehydes by reduction of their oximes.

In a typical analytical procedure the alkene in pentane solution at −65 to −75° is mixed with a pentane solution of ozone (0.03 M) at the same temperature. Reaction occurs in a few minutes and the ozonide is decomposed at 0° with hydrogen and Lindlar's catalyst or in some other appropriate fashion. In another process ozonisation is effected at 0° in methanol solution and the alkoxy hydroperoxide reduced at 20–25° with zinc–acetic acid or hydrogen–palladium charcoal. These methods can be applied on a micro or macro scale and may be applied to acids or esters.

The ozonolysis of oleic acid to furnish azelaic acid (nonanedioic) is effected on an industrial scale and ozonolysis of other unsaturated acids or glycerides has furnished a range of interesting difunctional compounds as illustrated in the following equations.

$$RCH{=\!=}CH(CH_2)_nCOOH \longrightarrow HOOC(CH_2)_nCOOH \quad (n = 7, 8, 10, 11, 13)$$

11:1 10*e*, 18:1 9*c*, 22:1 13*c*

18:1 (9*c* and 12*c* and 15*c*)

$$RCH{=\!=}CH(CH_2)_nCOOCH_3 \longrightarrow X(CH_2)_nCOOCH_3$$

$$X = CHO \text{ or } CH_2OH$$

$$CH_3(CH_2)_{10}CH{=\!=}CH(CH_2)_4COOH \longrightarrow CH_3(CH_2)_{11}NH_2 + H_2N(CH_2)_5COOH$$

REFERENCES

1. E. N. FRANKEL, *Fatty Acids* (ed. E. H. Pryde), American Oil Chemists' Society, Champaign, 1979, p. 353.
2. E. N. FRANKEL, *Progress in Lipid Chemistry*, 1980, **19,** 1.
3. G. A. VELDINK, J. F. G. VLIEGENTHART, and J. BOLDINGH, *Progress in the Chemistry of Fats and Other Lipids*, 1979, **15,** 131.
4. J. F. G. VLIEGENTHART, *Chem. and Indy.* 1979, 241.
5. D. SWERN, *Fatty Acids* (ed. E. H. Pryde), American Oil Chemists' Society, Champaign, 1979, p. 236.
6. F. D. GUNSTONE, *Fatty Acids* (ed. E. H. Pryde), American Oil Chemists' Society, Champaign, 1979, p. 379.
7. E. H. PRYDE and J. C. COWAN, *Topics in Lipid Chemistry*, 1971, **2,** 1.

CHAPTER 8

Other Reactions of Double Bonds

A. HALOGENATION

Halogenation of unsaturated fatty acids has been extensively studied in connection with the determination of iodine value and with the bromination reaction.

The iodine value indicates the weight of iodine which adds to 100 g of material—whether acid, ester, or neutral lipid—and is a measure of the mean unsaturation of the sample. Though expressed in terms of iodine, the analytical procedure most often uses a solution of iodine monochloride in acetic acid (Wijs' reagent). Methyl oleate (85.6), linoleate (173.2), and linolenate (260.3) have the iodine values indicated in parenthesis. Now that fatty acid composition can be easily determined by gas-liquid chromatography, iodine values are less commonly measured.

Halogenation—particularly bromination—is usually a polar addition occurring in a *trans* manner. Bromination of oleic (9*c*) and elaidic (9*t*) acid give the *threo* (m.p. 28.5°) and *erythro* (m.p. 29.5–30°) isomers of 9,10-dibromostearic acid respectively. Linoleic acid furnishes two tetrabromides: a solid product (m.p. 115°, 9*R**, 10*R**, 12*R**, 13*R** racemate) and a liquid isomer (9*R**, 10*R**, 12*S**, 13*S** racemate). Linelaidic acid (9*t* 12*t*) gives a solid tetrabromide (m.p. 78°) and linolenic acid gives a solid hexabromide (m.p. 185°). The chlorides obtained from oleic, linoleic, and linolenic acid melt at 37°, 123°, and 189° respectively.

vic-Dibromo acids are debrominated by reaction with zinc or sodium iodide. Both are *trans*-elimination processes with the latter the more stereospecific. The alkene has the same configuration as that from which the dibromide was prepared, and bromination–debromination has been employed as a method of protecting double bonds.

$$\underset{\substack{\textit{cis}\\ \textit{trans}}}{-CH{=}CH-} \xrightarrow{Br_2} \underset{\substack{\textit{threo}\\ \textit{erythro}}}{-CHBrCHBr-} \xrightarrow{NaI} \underset{\substack{\textit{cis}\\ \textit{trans}}}{-CH{=}CH-}$$

B. METATHESIS[1]*

The olefin metathesis reaction is an equilibrium process which may be represented by two equations depending on whether the starting material is an unsymmetrical alkene or two different unsymmetrical alkenes:

* Superscript numbers refer to References at end of Chapter.

$$ACH{=}CHB \underset{}{\overset{\text{catalyst}}{\rightleftharpoons}} ACH{=}CHA + BCH{=}CHB$$

$$\left.\begin{array}{l} ACH{=}CHB \\ XCH{=}CHY \end{array}\right\} \overset{\text{catalyst}}{\rightleftharpoons} \left\{\begin{array}{l} ACH{=}CHX + ACH{=}CHA \\ ACH{=}CHY + BCH{=}CHB \\ BCH{=}CHX + XCH{=}CHX \\ BCH{=}CHY + YCH{=}CHY \end{array}\right.$$

These equilibria are established with a homogeneous or heterogeneous catalyst. The former is usually a mixture of a transition metal compound (e.g. tungsten hexachloride) with an organometallic compound (ethylaluminium chloride, tetramethyltin, triethylboron) or a Lewis acid. The heterogeneous catalyst may be molybdenum or tungsten oxide on alumina or silica. Under appropriate reaction conditions methyl oleate gives a mixture of hydrocarbon and diester, whilst a mixture of methyl oleate and hex-3-ene gives diester (18:1), monoester (12:1 Δ9), and hydrocarbons (18:1 and 12:1):

$$CH_3(CH_2)_7CH{=}CH(CH_2)_7COOCH_3 \rightleftharpoons \left\{\begin{array}{l} CH_3(CH_2)_7CH{=}CH(CH_2)_7CH_3 \\ CH_3OCO(CH_2)_7CH{=}CH(CH_2)_7COOCH_3 \end{array}\right.$$

$$\left.\begin{array}{l} CH_3(CH_2)_7CH{=}CH(CH_2)_7COOCH_3 \\ CH_3CH_2CH{=}CHCH_2CH_3 \end{array}\right\} \rightleftharpoons \left\{\begin{array}{l} CH_3OCO(CH_2)_7CH{=}CH(CH_2)_7COOCH_3 \\ CH_3CH_2CH{=}CH(CH_2)_7COOCH_3 \\ CH_3(CH_2)_7CH{=}CH(CH_2)_7CH_3 \\ CH_3(CH_2)_7CH{=}CHCH_2CH_3 \end{array}\right.$$

Methyl linoleate gives a range of alkenes (12:1, 15:2, 18:3, 21:4, 24:5), monoesters of the 15:1, 18:2, 21:3, 24:4, 27:5 acids, and diesters of the 18:1, 21:2, 24:3, 27:4, 30:5 acids.

C. STEREOMUTATION, DOUBLE BOND MIGRATION, CYCLISATION[2]

Olefinic compounds can change to isomeric forms by stereomutation (change of configuration of the double bond), double bond migration, or cyclisation. Sometimes these changes occur inadvertently—for example during processing—and it is important to know when this is happening: sometimes they are promoted with appropriate reagents.

Steam deodorisation of linolenic acid at about 230° produces a number of configurational isomers among which the 9*t*12*c*15*c* and 9*c*12*c*15*t* trienes predominate and the 9*t*12*c*15*t* and 9*t*12*t*15*c* isomers are minor products. Under the same conditions linoleic acid gives some of the 9*t*12*c* and 9*c*12*t* dienes.

Cis and *trans* isomers can be interconverted by a sequence of stereospecific reactions which add up to stereomutation. A single product is usually obtained as in the sequence:

$$\underset{cis}{-CH{=}CH-} \xrightarrow{ArCO_3H} \underset{cis}{-\overset{O}{\overbrace{CHCH}}-} \xrightarrow{LiPPh_2} \underset{threo}{-CH(OH)CH(OPPh_2)-} \xrightarrow{MeI} \underset{trans}{-CH{=}CH-}$$

It is more usual, however, to treat the readily available *cis* isomer with some reagent which sets up a *cis/trans* equilibrium and then to isolate the *trans* isomer from the reaction mixture by crystallisation or by silver ion chromatography. The reagent selected should promote stereomutation without double bond migration or hydrogen transfer. Reaction with selenium at high temperature is out of favour because double bond migration also occurs, but a mixture of oxides of nitrogen (mainly NO_2), produced by interaction of sodium nitrite and nitric acid, or sulphur compounds such as 3-mercaptopropionic acid or an aromatic sulphinic acid may be used.

In monoenes and non-conjugated polyenes, each *cis* double bond is converted to the *trans* form in 75–80 % yield. Conjugated polyenes isomerise more readily and give a higher proportion of the all-*trans* isomer. This change may be effected by exposure to light and iodine.

When treated with strong base a double bond may migrate through the sequence of events shown in the equation:

$$-CH{=}CHCH_2- \rightleftharpoons -CH{=}CH\bar{C}H- \longleftrightarrow -\bar{C}HCH{=}CH- \rightleftharpoons -CH_2CH{=}CH-$$

Reaction occurs via a resonance-stabilised carbanion and double bond migration is usually accompanied by stereomutation. With methylene-interrupted polyenes the reaction occurs under slightly milder conditions and leads to compounds with conjugated unsaturation. Linoleic acid, for example, gives the 9*c*11*t* and 10*t*12*c* octadecadienoic acids. Linolenic acid gives, initially, a mixture of trienes with diene or triene conjugations.

Alkali isomerisation of polyenes with three or more double bonds also produces cyclised compounds. These usually contain monocyclic (cyclohexadiene) or bicyclic (indene) systems which then undergo hydrogen exchange to furnish saturated alicyclic and aromatic systems. The indene system is thought to result from an internal Diels–Alder reaction of a Δ1,3,8-triene unit.

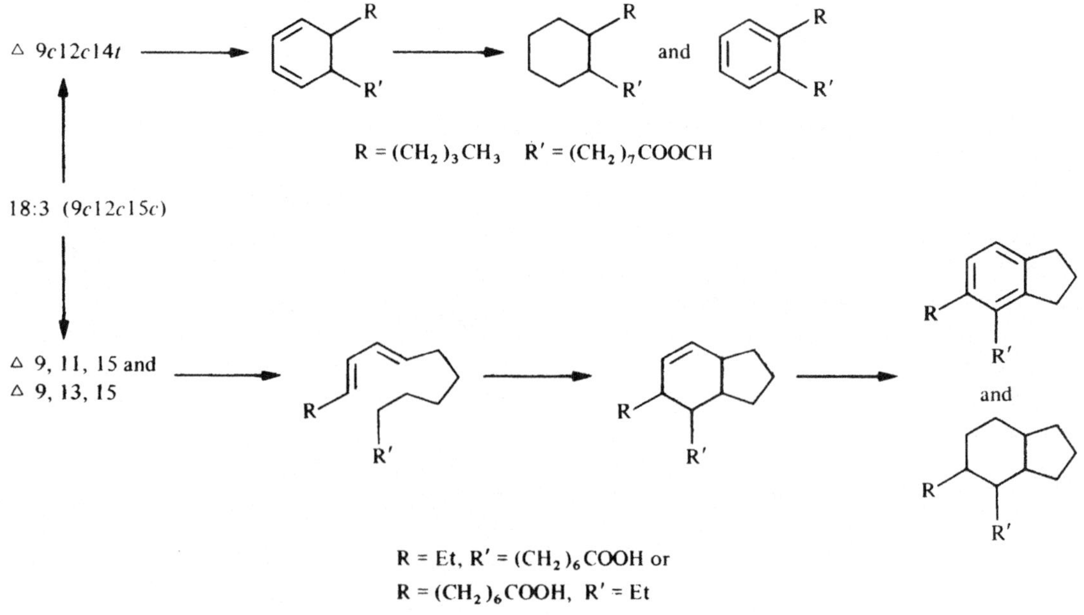

Similar cyclic compounds are produced from tung oil which is rich in the already conjugated 18:3 (9*c*11*t*13*t*) acid. Alkali is not required, but the thermal reaction is facilitated by sulphur which, presumably, promotes *cis* to *trans* isomerisation.

Acid-catalysed double bond migration converts oleic acid to the γ-lactone of stearic acid. This reaction involves reversible protonation-deprotonation of the double bond with trapping of the C-4 carbonium ion by the carboxyl group to produce a lactone.

D. DIMERISATION[3]

Derivatives of dimeric acids which remain liquid at low temperatures have been used as surface-active agents, corrosion inhibitors, and as a source of improved alkyd resins. Dimerisation of fatty acids occurs in the presence of radical sources, on heating, and under the influence of clay catalysts. The last of these is employed on a commercial basis.

In the clay-catalysed reactions even oleate will furnish a cyclic dimer along with a mixture of saturated and unsaturated (mainly *trans*) straight-chain and branched-chain C_{18} compounds formed by hydrogen transfer and rearrangement. Dimerisation results from diene synthesis between a conjugated diene (produced from monoene by hydrogen transfer) and a monoene. By further hydrogen transfer, cyclohexene derivatives are converted to cyclohexane and benzene compounds. The monocyclic dimer is accompanied by acyclic and bicyclic dimers such as the compounds (**2**) and (**3**). Linoleate reacts in a similar manner.

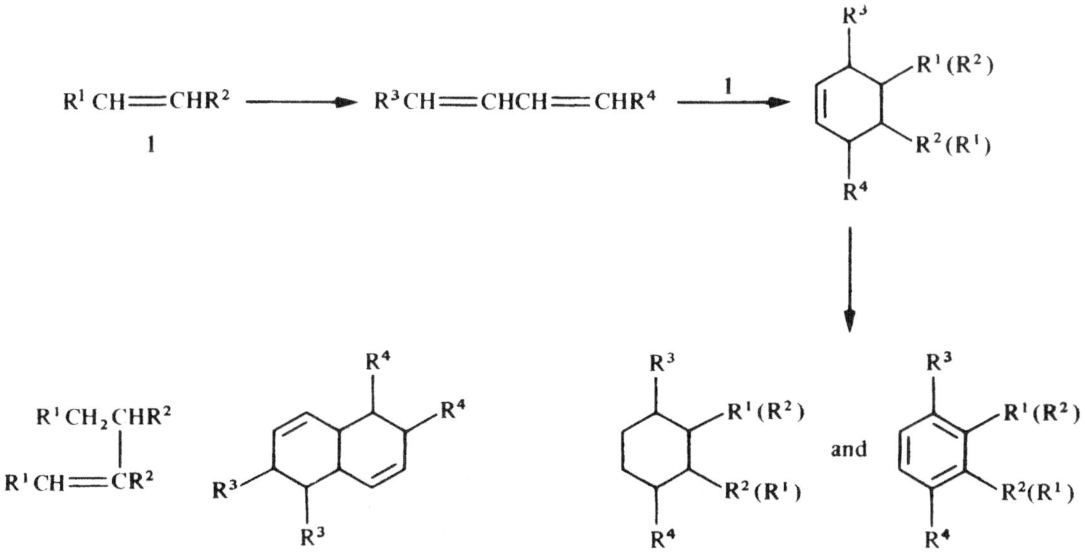

2 3

Monoene acids give mainly acyclic and a monocyclic dimer, but diene acids give mainly mono- and bicyclic dimers. Tall oil fatty acids are commonly used for dimerisation.

E. OTHER DOUBLE BOND REACTIONS

(i) Alkenes react with mercuric acetate (or other mercury salts) and methanol (or other nucleophiles such as alcohols or carboxylic acids) to give a methoxy mercuriacetate adduct in a *trans* addition process. This intermediate regenerates the alkene in its original stereoisomeric form on treatment with hydrochloric acid. Alternatively the mercury group is replaceable by bromine or by hydrogen as shown in the following equations:

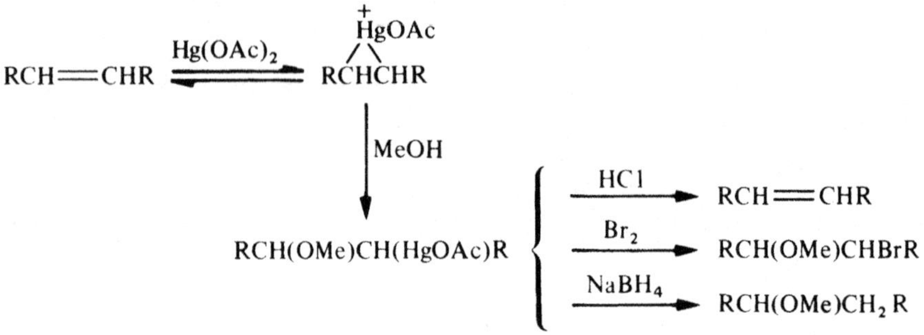

Intramolecular reaction occurs with olefinic compounds containing an appropriately placed nucleophilic group. This is illustrated in the reaction of arachidonyl alcohol:

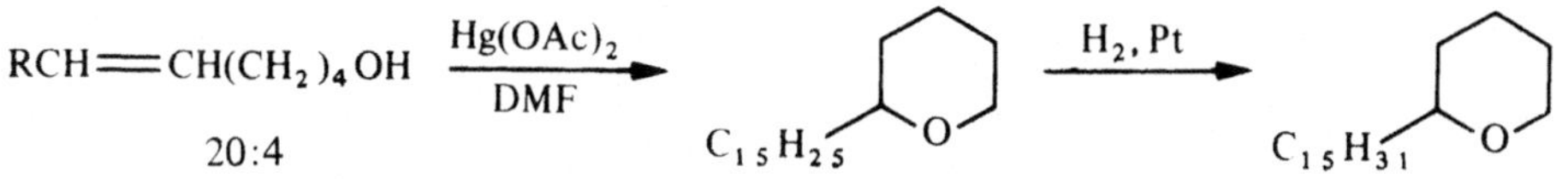

(ii) Cyclopropane compounds are readily obtained from alkenes by the Simmons–Smith reaction which, in its original form, involves reaction with methylene iodide and zinc–copper couple.[4]

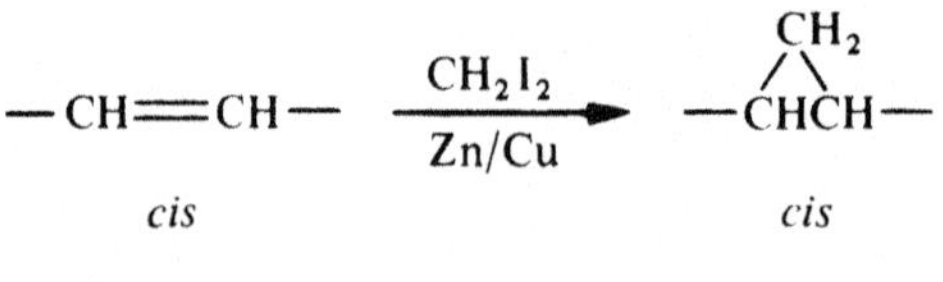

(iii) In acid-catalysed addition reactions the new substituent is seldom confined to the unsaturated carbon atoms, the reaction being accompanied or preceded by double bond migration and stereomutation. Thus the simple Friedel–Crafts reaction between oleic acid and benzene in the presence of aluminium trichloride gives a mixture of the 5- to 17-phenylstearic acids:

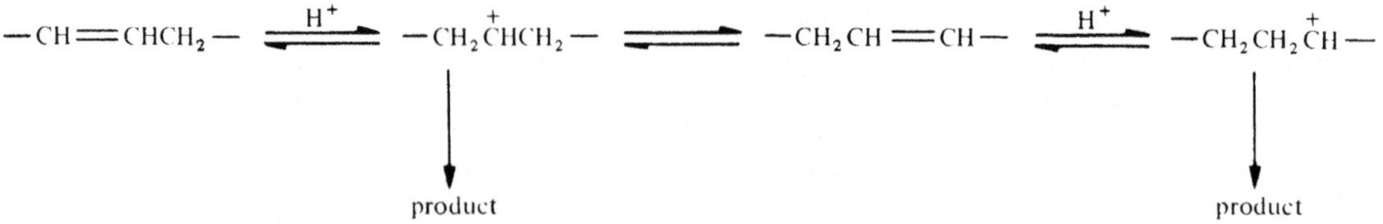

(iv) Reaction involving addition of carbon monoxide to an alkene yields a product with an additional formyl or related group.[1] Best known is the hydroformylation (oxo) reaction between an alkene such as oleic acid, synthesis gas ($H_2 + CO$) under pressure, and cobalt carbonyl. Reaction at 100° gives a mixture of formyl- and hydroxymethyl-stearic acids. With a rhodium catalyst and triphenylphosphine, isomerisation is insignificant and the product is largely confined to 9- and 10-formylstearic acids. The product is easily converted to the range of compounds shown in the following equations:

$$-CH{=}CH- \xrightarrow[Co(CO)_8]{H_2,\,CO} -CH_2CH(CHO)- \left\{ \begin{array}{l} \longrightarrow -CH_2CH(CH_2OH)- \\ \longrightarrow -CH_2CH[CH(OMe)_2] \\ \longrightarrow -CH_2CH(COOH)- \end{array} \right.$$

REFERENCES

1. E. N. Frankel, *Fatty Acids* (ed. E. H. Pryde), American Oil Chemists' Society, Champaign, 1979, p. 426.
2. R. W. Johnson and E. H. Pryde, *Fatty Acids* (ed. E. H. Pryde), American Oil Chemists' Society, Champaign, 1979, p. 319.
3. R. W. Johnson, *Fatty Acids* (ed. E. M. Pryde), American Oil Chemists' Society, Champaign, 1979, p. 343.
4. W. W. Christie, *Topics in Lipid Chemistry*, 1970, **1**, 1.

CHAPTER 9

Reactions of the Carboxyl Group

A. HYDROLYSIS

Ester or lipid hydrolysis is most conveniently effected in the laboratory by aqueous ethanolic alkali. Acidification of the hydrolysate liberates fatty acids which can then be extracted with hexane or other organic solvent. Non-acidic compounds (unsaponifiables) such as hydrocarbons, long-chain alcohols, sterols, and glycerol ethers will also be present in the organic extract, but glycerol remains in the aqueous phase.

The production of soap by alkaline hydrolysis (saponification) is usually carried out around 100°, with glycerol being recovered as a second commercial product. Sodium and potassium salts are used as soaps. Salts of other metals are used to promote polymerisation of drying oils, in the manufacture of grease and lubricants, an as ingredients in plastics formulations.

Fats can also be hydrolysed to free acids in what is probably a homogeneous reaction between fat and water dissolved in the oil phase (fat splitting). This process occurs in the presence of sulphonated long-chain alkylbenzenes (24–48 hr, ~100°) or of oxides of zinc or magnesium or calcium (2–3 hr, 250°, 700 psi).

Partial hydrolysis of fats occurs during digestion under the influence of lipases (Chapter 4). Lipases present in seeds may also promote some hydrolysis, so that most extracted lipids contain some free acid and some partial glycerides. This is an undesirable change, since removal of free acid during processing is accompanied by some loss of fat.

B. ESTERIFICATION, ALCOHOLYSIS, ACIDOLYSIS, INTERESTERIFICATION

Esters can be prepared by interaction of a carboxylic acid or an acyl derivative with an alcohol or its equivalent (esterification) or by reaction of an ester with an alcohol (alcoholysis), an acid (acidolysis), or another ester (interesterification). Each process usually requires an acidic or basic catalyst.

1. Esterification

Fatty acids are frequently studied in the form of their esters (e.g. for chromatographic purposes) and most commonly as methyl esters. Methyl esters are made from acids by reaction with methanol containing sulphuric acid (1–2%), hydrogen chloride (5%), or boron trifluoride (12–14%) as catalyst, or by reaction with diazomethane. Acid-catalysed procedures are not suitable for acids containing cyclopropane, cyclopropene, epoxide, or allylic hydroxy groups, nor for conjugated polyene acids. Such acids, however, can be safely esterified with diazomethane or the ester can be obtained directly from lipid by base-catalysed transesterification.

2. Alcoholysis

Alcoholysis is widely employed to convert lipids directly to methyl esters without first isolating the free acids and to prepare partial glycerides through interaction of a triacylglycerol and free glycerol. For methanolysis the fat or oil is dissolved in excess of methanol containing a catalyst

$$R^1COOR^2 + R^3OH \overset{\text{catalyst}}{\rightleftharpoons} R^1COOR^3 + R^2OH$$

with benzene, toluene, or dichloromethane as co-solvent. The catalyst may be acidic (sulphuric acid, hydrogen chloride, boron trifluoride) or basic (sodium methoxide).

Acidic methanolysis:

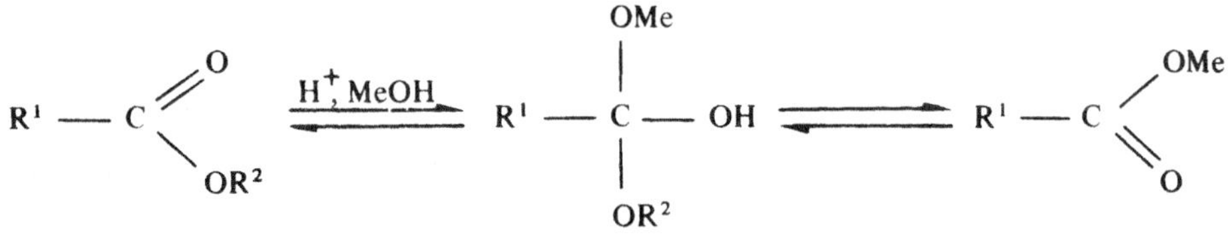

Basic methanolysis:

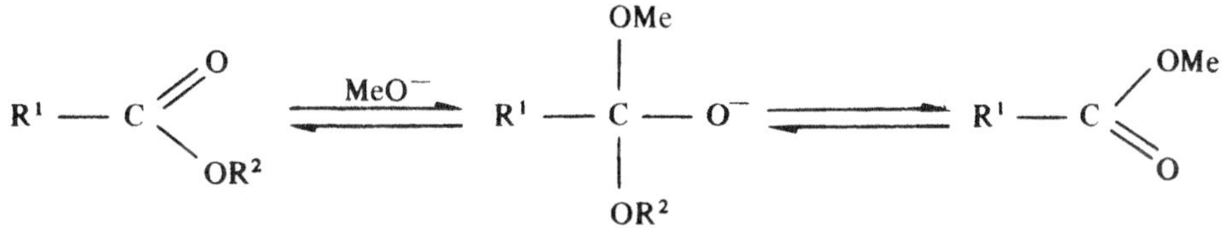

When a triacylglycerol and glycerol are heated together in the presence of sodium hydroxide or sodium methoxide the equilibrium shown in the following equation is established:

$$\text{triacylglycerol} + \text{glycerol} \rightleftharpoons \text{monoacylglycerol} + \text{diacylglycerol}$$

This is an important method of preparing mono- and diacylglycerols. The composition of the equilibrium mixture of all four components depends on the relative amount of triacylglycerol and of glycerol dissolved in the lipid phase.

3. Acidolysis

This less common process involves the interaction of ester and carboxylic acid in the presence of sulphuric acid or zinc oxide or calcium oxide. Applied to natural glycerides and lauric acid, for example, C_{16} and C_{18} acids are replaced by the C_{12} acid.

$$R^1COOR^2 + R^3COOH \rightleftharpoons R^3COOR^2 + R^1COOH$$

4. Interesterification*

A natural fat or oil is a mixture of triacylglycerols in which the acyl groups are distributed in a non-random manner (Chapter 3). Under the influence of a basic catalyst (sodium hydroxide, sodium methoxide, or a sodium–potassium alloy) at about 80° the acyl groups are redistributed both intermolecularly and intramolecularly until a wholly random distribution is finally achieved. This redistribution of acyl groups leads to a change in the physical properties of the triacylglycerol mixture. Thus the melting point of soybean oil is raised from −7° to +6° and that of cottonseed oil from 10° to 34°.

This procedure can also be applied to mixtures of oils and fats. All the acids in the two (or more) oils will then be distributed at random, and this provides a method of transferring saturated fatty acids to predominantly unsaturated glycerides and *vice versa.*

These changes can be further modified if the interesterification process is carried out at a lower temperature (0–40°), so that fully saturated glycerides (S_3) crystallise from the reaction mixture. This is called directed interesterification. Crystallisation disturbs the equilibrium in the liquid phase so that more S_3 forms and again separates producing, finally, more S_3 and U_3 and less of the mixed glycerides. This usually causes an increase in the melting range of the fat.

$$U_3 + S_3 \rightleftharpoons US_2 + U_2S$$

Interesterification procedures are used industrially to improve the physical properties of lard, to produce cocoa butter substitutes from cheaper oils, and to produce fats containing acetic acid.

The isomerisation of mono- and diacylglycerols depends on a similar acyl migration. 1- and 2-Monoacylglycerols isomerise to a 90:10 mixture of the two isomers under acidic, basic, or thermal conditions. Such a change may occur during chromatography, though this is reported to be less serious if the adsorbent is impregnated with boric acid.

Diacylglycerols behave similarly. The equilibrium mixture of 1,2- (40 %) and 1,3- (60 %) isomers varies somewhat with temperature.

C. ACID CHLORIDES AND ANHYDRIDES[1†]

Acid chlorides are made from fatty acids by reaction with phosphorus trichloride, phosphorus pentachloride, phosphorus oxychloride, phosgene, oxalyl chloride, thionyl chloride, or triphenylphosphine and carbon tetrachloride. Phosphorus trichloride is probably the most economical reagent for large-scale reaction. Improved yields have been claimed when the reaction is carried out in absence of oxygen. The most common laboratory procedures require the acid to stand at room temperature for 3–5 days with thionyl chloride (2.4 mol) for saturated acids and with oxalyl chloride (1.8 mol) for unsaturated acids.

Acid anhydrides are prepared by interaction of fatty acids or, better, acid chlorides with acetic anhydride at reflux temperature. The least volatile component of the equilibrium mixture (acetic acid or acetyl chloride) is continuously removed by distillation to shift the equilibrium in the desired direction. Dicyclohexylcarbodi-imide will dehydrate acids at room temperature to give the anhydride

* See also Chapter 17.

† Superscript numbers refer to References at end of Chapter.

$$(CH_3CO)_2O \underset{}{\overset{RCOOH}{\rightleftharpoons}} RCOOCOCH_3 \overset{RCOOH}{\rightleftharpoons} (RCO)_2O$$

$$CH_3 + COOH \qquad CH_3 + COOH$$

$$2RCOCl + (CH_3CO)_2O \rightleftharpoons (RCO)_2O + 2CH_3COCl$$

$$2RCOOH + C_6H_{11}N{=}C{=}NC_6H_{11} \longrightarrow (RCO)_2O + C_6H_{11}NHCONHC_6H_{11}$$

Both the acid chlorides and anhydrides are effective acylating agents used in the synthesis of acylglycerols and phospholipids (Chapter 10).

Mixed anhydrides with methanesulphonic or toluene-4-sulphonic acid are also reported to be effective acylating agents. They are made as indicated in the following equations:

$$RCOOC(CH_3){=}CH_2 + CH_3OSO_3H \longrightarrow RCOOSO_2CH_3 + (CH_3)_2C{=}O$$

$$RCOCl + AgOSO_2C_6H_4CH_3 \longrightarrow RCOOSO_2C_6H_4CH_3 + AgCl$$

D. NITROGEN- AND SULPHUR-CONTAINING DERIVATIVES

Long-chain nitrogen-containing compounds are extensively used in plastics formulations and as lubricants, detergents, and froth flotation agents. Many of these are made *via* amines prepared from fatty acids to the extent of 200,000 tons per annum.

$$RCOOH \longrightarrow RCONH_2 \longrightarrow RCN \longrightarrow RCH_2NH_2$$

$$RCOOH \longrightarrow RCN$$

Amides are prepared industrially by reaction of fatty acid with ammonia (or other amine) at about 200°. The *nitriles* are produced by thermal decomposition of the amides or directly from acids and ammonia in the presence of a dehydrating catalyst such as alumina. *Amines* (RCH_2NH_2 and RCH_2NHCH_2R) are then obtained from the nitrile by catalytic reduction with Raney nickel in the presence of ammonia. The long-chain amines are most commonly used as *tertiary amines* (RCH_2NMe_2, $(RCH_2)_2NMe$, $RCH_2N[(CH_2CH_2O)_nH]_2$) or as quaternary ammonium compounds ($RCH_2\overset{+}{N}Me_3\overset{-}{Cl}$, $(RCH_2)_2\overset{+}{N}Me_2\overset{-}{Cl}$).

Long-chain alcohols form sulphates ($ROSO_3H$) by reaction with sulphuric acid, sulphur trioxide, or sulphamic acid, but the related methane- or 4-toluene-sulphonate ($ROSO_2R'$, $R' = CH_3$ or $4\text{-}CH_3C_6H_4$) are usually more convenient compounds. Sulphonate anions are good leaving groups and these compounds are frequently used as alkylating agents.

$$ROH \xrightarrow{i} \left\{\begin{matrix} ROMs \\ ROTs \end{matrix}\right\} \xrightarrow{ii} RX$$

i, $CH_3SO_2Cl(MsCl)$ or $4-CH_3C_6H_4SO_2Cl(TsCl)$;
ii, $LiAlH_4(X = H)$ or $MgBr_2(X = Br)$ or $KCN(X = CN)$
or $NaCH(CO_2Et)_2$ $(X = CH(CO_2Et)_2$ or $H_2O_2(X = OOH)$
or $R'OH(X = OR')$ or $R'COOH(X = OCOR')$

Sulphation of saturated acids or esters to give compounds of structure $RCH(SO_3H)COOR'$ ($R' = H$ or CH_3) is effected by sulphur trioxide.

E. α-ANIONS OF CARBOXYLIC ACIDS[2]

Saturated and monoene carboxylic acids react with lithium isopropylamide ($Pr^i{}_2NLi$) in tetrahydrofuran-hexamethylphosphoramide to form an α-dianion ($R\bar{C}HCO\bar{O}$). These are useful intermediates for the preparation of α-substituted acids (RCHYCOOH) or compounds of the type RCH_2Y when the reaction is accompanied by decarboxylation. The following are typical of the compounds which have been prepared using the reagent indicated in parentheses: RCH(R′)COOH (R′Br), $RCH(CH_2OH)COOH$ and $RC(=CH_2)COOH$ (CH_2O), RCH(OOH)-COOH and RCH(OH)COOH (O_2), $RCHICOOH(I_2)$, RCHBrCOOH (Br_2), $RCH(COOH)_2$ [CO_2 or ClCOOEt or $(R'O)_2CO$], RCH(COR′)COOH (R′COCl), RCH_2CHO via RCH(CHO)-COOH (HCOOEt), and $RCH_2NO_2(PrONO_2)$.

F. PEROXY ACIDS[3]

Organic peroxy acids of general formula RCO_3H are intramolecular hydrogen-bonded monomeric compounds in contrast to carboxylic acids which are dimeric in the solid state and in solution. The C_{16} and C_{18} peroxy acids melt at 61° and 65° respectively.

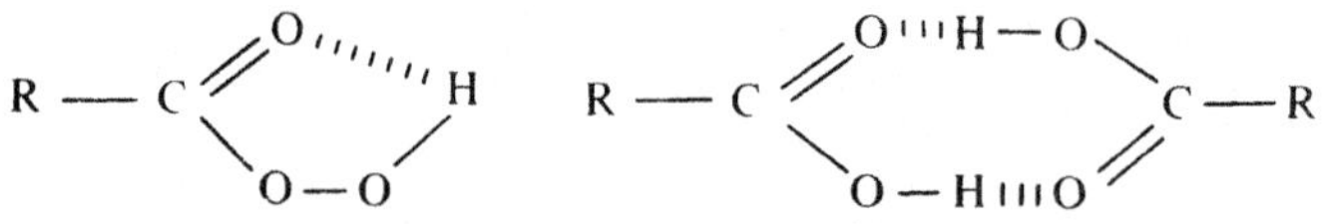

peroxy acid monomer   carboxylic acid dimer

They are formed in an equilibrium reaction between carboxylic acids and hydrogen peroxide (30–98%), usually in the presence of an acidic catalyst (sulphuric acid, a sulphonic acid, a sulphonic acid ion exchange resin)

$$RCO_2H + H_2O_2 \rightleftharpoons RCO_3H + H_2O$$

Peroxy acids undergo thermal decomposition by radical and non-radical processes to give alcohols and carboxylic acids respectively. Their other chemical reactions reflect their effectiveness

$$RCO_2H + \tfrac{1}{2}O_2 \xleftarrow{\text{non - radical}} RCO_3H \xrightarrow{\text{radical}} ROH + CO_2$$

as oxidising agents. They oxidise alkenes to epoxides and diols (Chapter 7), ketones to esters or lactones, thioethers to sulphoxides and sulphones, and tertiary amines to amine oxides thus:

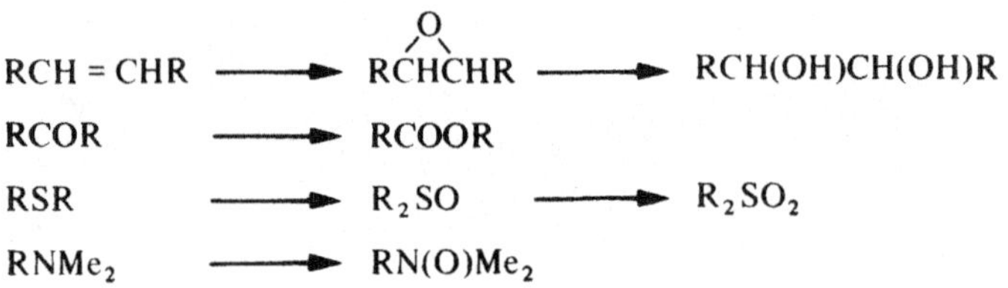

REFERENCES

1. R. A. GRIMM, *Fatty Acids* (ed. E. H. Pryde), American Oil Chemists' Society, Champaign, 1979, p. 218.
2. L. S. SILBERT and P. E. PFEFFER, *Fatty Acids* (ed. E. H. Pryde), American Oil Chemists' Society, Champaign, 1979, p. 272.
3. D. SWERN, *Fatty Acids* (ed. E. H. Pryde), American Oil Chemists' Society, Champaign, 1979, p. 236.

CHAPTER 10

Synthesis

A. WHY SYNTHESIS?

Much effort has been expended in developing procedures for synthesising long-chain acids and lipids. Why is this necessary when these occur so widely?

Whilst common natural acids such as oleic and linoleic are readily isolated from natural sources (Chapter 2), other acids—particularly the unsaturated members—occur so rarely or are so difficult to purify from the natural mixtures in which they occur that it is often easier to synthesise them. Additionally, there is sometimes a need for acids that do not occur naturally. For example, isomers of oleic acid such as the $\Delta 8c/t$ or $\Delta 10c/t$ acids present in partially hydrogenated fat may be required for a detailed examination of their physical, chemical, or biological properties. These are best obtained by synthesis. For research purposes there is also a demand for isotopically labelled acids, and these may have to be prepared by full chemical synthesis.

The most commonly used synthetic routes involve acetylenic compounds or the Wittig reaction. In addition, procedures for converting natural or synthetic acids to their higher homologues are also useful. These include reaction sequences involving the nitrile (chain extension by one carbon atom), malonic ester (two carbon atoms), enamine synthesis (five or six carbon atoms), and anodic synthesis (several carbon atoms).

As already indicated, natural lipids are almost always complex mixtures. Whilst it is possible to separate individual lipid classes the separation of individual members within a class is more difficult. Though sometimes possible on an analytical scale using the most modern chromatographic procedures, for most purposes pure individual lipids can only be obtained in useful quantities by synthetic means. The acids required for this purpose and some other subunits of the final structure may themselves be obtained from natural sources, but specific assembly procedures are required.

B. SYNTHESIS OF ACIDS

1. The acetylenic route to *cis*-mono- and poly-enoic acids[1,2]*

The success of this approach depends on the ease with which acetylenic compounds can be alkylated and on our ability to reduce poly-ynes to all-*cis* polyenes. The procedures can usually be modified to give acids with mixed types of unsaturation. The acetylenic units become the olefinic

* Superscript numbers refer to References at end of Chapter.

centres. The required carboxyl group may be introduced as such, by modification of a halide *via* cyanide or malonate, or by oxidation of an alcohol.

Monoynoic acids are prepared by alkylation of acetylene with an alkyl halide or its equivalent and either an $\alpha\omega$-chloroiodoalkane (exploiting the differing reactivity of the two carbon–halogen bonds) or an ω-halogeno acid or alcohol.

$$HC{\equiv}CH \xrightarrow{\text{i}} RC{\equiv}CH \xrightarrow{\text{ii}} RC{\equiv}C(CH_2)_nCl \begin{cases} \xrightarrow{\text{iii}} RC{\equiv}C(CH_2)_nCOOMe \\ \xrightarrow{\text{iv}} RC{\equiv}C(CH_2)_{n+1}COOMe \end{cases}$$

i, $NaNH_2$, RBr; ii, $NaNH_2$, $I(CH_2)_nCl$; iii, KCN; MeOH, HCl;
iv, $CH_2(CO_2Et)_2$, NaOEt; MeOH, HCl

Most of the important polyunsaturated acids have a single methylene group between adjacent unsaturated centres and it is for this reason that propargylic ($HC{\equiv}CCH_2X$) and sometimes allylic compounds ($H_2C{=}CHCH_2X$) are widely used. A key reaction is formulated below. It involves a

$$R^1C{\equiv}CCH_2X + BrMgC{\equiv}CR^2 \xrightarrow[\text{THF}]{Cu^I} R^1C{\equiv}CCH_2C{\equiv}CR^2$$
(X= I, OMs, OTs)

propargylic (or allylic) halide and an ethynyl compound as its magnesium bromide derivative in the presence of Cu^I ions. R^1 or R^2 must contain a functional group which is or will become the carboxyl function. Care must be taken to select intermediates which give the best overall yield. Kunau claims that under optimum conditions pentaenes and heptaenes can be obtained in 20% and 10% yield respectively. In the scheme set out below by way of illustration the pentaynoic acid is obtained from an alkyl halide ($CH_3(CH_2)_mBr$), propargyl alcohol (4 mols), and the acetylenic alcohol $HC{\equiv}C(CH_2)_nCH_2OH$, itself made from acetylene and a halogeno alcohol.

$$CH_3(CH_2)_mBr \xrightarrow{\text{i}} CH_3(CH_2)_mC{\equiv}CCH_2OH^{\dagger} \xrightarrow{\text{ii}}$$
$$CH_3(CH_2)_m(C{\equiv}CCH_2)_3OH^{\dagger} \xrightarrow{\text{iii}} CH_3(CH_2)_m(C{\equiv}CCH_2)_4C{\equiv}C(CH_2)_nCOOH$$

i $BrMgC{\equiv}CCH_2Othp^*$, H_3O^+; ii, $BrMgC{\equiv}CCH_2C{\equiv}CCH_2Othp^*$, H_3O^+;
iii, $BrMgC{\equiv}CCH_2C{\equiv}C(CH_2)_nCOOMgBr^*$, H_3O^+

thp = tetrahydropyranyl

† the alcohols react as their iodides or mesylates or tosylates

*prepared from propargyl alcohol.

The poly-yne acids are solids which can be preserved at low temperatures. They are purified by crystallisation prior to partial hydrogenation with hydrogen and Lindlar's catalyst (palladium on calcium carbonate containing some lead) in the presence of quinoline to enhance selectivity. Under these conditions *cis* olefinic compounds are obtained. The polyene is finally purified from compounds which have been over-reduced or contain *trans* double bonds by silver ion chromatography.

2. The Wittig reaction

The Wittig reaction is a means of obtaining alkenes by reacting aldehydes (or ketones) with a phosphorane derived from an alkyl halide. Stabilised phosphoranes such as those with a conjugated double bond produce *trans* alkenes, while non-stabilised phosphoranes give *cis* double bonds. In the presence of certain added salts *trans* alkenes can also be obtained from non-stabilised phosphoranes.

$$R^1CHO + BrCH_2R^2 \xrightarrow{Ph_3P,\ base} R^1CH{=}CHR^2$$

The alkyl halide R^2CH_2Br is the source of the phosphorane $R^2CH{=}PPh_3$ which can, however, be replaced by a phosphine oxide ($R^2CH_2POR_2$) or a phosphonate ($R^2CH_2PO(OEt)_2$). The conventional Wittig reaction is illustrated in the following synthesis of calendic (18:3 8*t*10*t*12*c*) and catalpic (18:3 9*t*11*t*13*c*) acids.

$$CH_3(CH_2)_mC{\equiv}CCH\overset{t}{=}CHCH_2OH \xrightarrow{i} CH_3(CH_2)_mC{\equiv}CCH{=}CHCH{=}PPh_3$$

$$\xrightarrow{ii} CH_3(CH_2)_mCH{=}CHCH{=}CHCH{=}CH(CH_2)_nCOOCH_3$$

calendic acid $n = 6$ $m = 4$
catalpic acid $n = 7$ $m = 3$

i. PBr_3; Ph_3P; NaOMe; ii, $OHC(CH_2)_nCOOCH_3$; H_2 Lindlar's catalyst

3. Isotopically labelled acids[3]

Isotopically labelled acids are used in studies of lipid metabolism, biosynthesis, and in mechanistic and kinetic studies. When the isotope is radioactive (^3H, ^{14}C) the labelled molecules are readily recognised with the help of a scintillation counter. Compounds containing stable isotopes (^2H, ^{13}C) are recognised by mass spectrometry or nmr spectroscopy. In general the isotopic element should be introduced into a synthesis as late as possible. Reference is made below to a few typical procedures for introducing isotopic labels. They are based on familiar reactions and labelled compounds, once obtained, can be incorporated into the synthetic sequences already described.

Reduction of carbon–carbon multiple bonds can be carried out with deuterium or tritium. As already discussed (Chapter 6), with a heterogeneous catalyst this is frequently a complex reaction proceeding with hydrogen exchange *via* isomeric intermediates with the result that many isotopic atoms may be introduced. Better results follow the use of a homogeneous catalyst such as tris(triphenylphosphine)rhodium I chloride. Lindlar's catalyst can be safely used so that alkenes with two ^2H or ^3H atoms can be obtained from acetylenic compounds. The safer route to saturated acids uses labelled hydrazine which effects reduction without double bond migration.

$$-C{\equiv}C- \xrightarrow{D_2} -CD{=}CD- \xrightarrow{D_2} -CD_2CD_2-$$

$$-CHDCHD- \xleftarrow{N_2D_4} -CH{=}CH \xrightarrow{N_2T_4} -CHTCHT-$$

A single isotopic hydrogen atom can be introduced starting with hydroxy compounds by modification of the lithium aluminium hydride reduction of a mesylate or tosylate.

$$>CHOH \xrightarrow{MsCl} >CHOMs \begin{cases} \xrightarrow{LiAlD_4} >CHD \\ \xrightarrow{LiAlT_4} >CHT \end{cases}$$

Useful deuterium-containing intermediates are made by base-catalysed hydrogen-deuterium exchange of protons α to a carboxyl group. These can then be involved in syntheses via acetylenic compounds or by the Wittig reaction.

$$RCH_2COOH \xrightarrow{NaOD,\ D_2O} RCHDCOOH \xrightarrow{NaOD,\ D_2O} RCD_2COOH$$

The carbon isotopes (^{13}C, ^{14}C) are generally incorporated through the use of labelled carbon dioxide (from labelled carbonate) or cyanide. These reactions can furnish end products (carboxyl labelled acids) or provide intermediates for further synthesis.

$$RX \xrightarrow{Mg,\ Et_2O} RMgX \xrightarrow{{}^*CO_2,\ H_3O^+} R\,{}^*COOH$$

$$RX \xrightarrow{K\,{}^*CN} R{}^*\,CN \xrightarrow{CH_3OH,\ H_3O^+} R\,{}^*COOCH_3$$

C. SYNTHESIS OF ACYLGLYCEROLS[4,5]

1. Introduction

The position of acyl groups attached to glycerol will be shown by numbers 1, 2, and 3 and the compounds will be racemic unless otherwise indicated by the prefix *sn*.

A major problem in the synthesis of pure acylglycerols is the ease with which acyl groups in mono- and diacylglycerols migrate from the oxygen to which they are attached to an adjacent free hydroxyl group. This transesterification process is catalysed by acid or base or heat. Both 1- and 2-monoacylglycerols readily furnish an equilibrium mixture containing 10% of the latter and 1,2- and 1,3-diacylglycerols form mixtures having 60–80% of the latter.

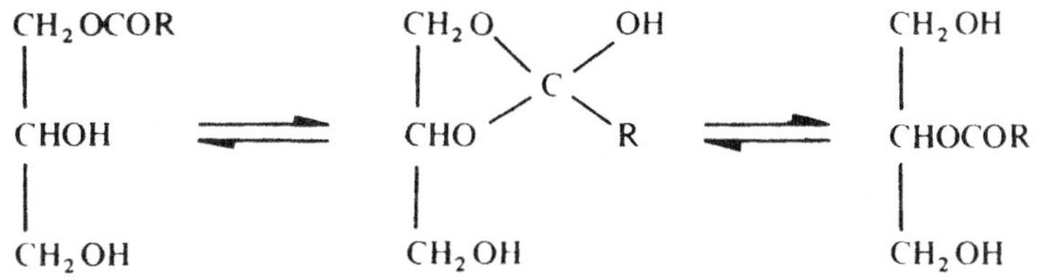

Procedures for obtaining acylglycerols with the desired acyl groups attached at the appropriate oxygen atom are based either on the use of protecting groups or on the greater reactivity of primary over secondary hydroxyl groups. It is essential in all work-up procedures to avoid conditions which promote acyl migration.

Purification of intermediate and final products is usually effected by careful crystallisation or chromatography. The purity of synthetic acylglycerols is checked by gas chromatography of the compound and of its component acids, by thin-layer chromatography, and by the enzymatic procedures already described.

2. Acylation procedures

Acylation is generally effected in chloroform solution by reaction at room temperature for 1–3 days or at 100° for 4 hr with a slight excess of acyl halide and an equivalent amount of pyridine or other tertiary base. The acid chlorides are prepared from pure acids by reaction with excess of thionyl chloride ($SOCl_2$) for saturated acids or oxalyl chloride (ClCOCOCl) for unsaturated acids. Alternative acylation procedures include reaction with a carboxylic acid and an acid catalyst (toluene-4-sulphonic acid, trifluoroacetic anhydride, sulphonated polystyrene resin) or with an acid anhydride and trifluoromethylsulphonic acid. The last reaction is reported to occur at room temperature in about 3 h.

3. Protecting group

In the presence of an acid catalyst such as toluene-4-sulphonic acid, glycerol reacts with acetone to form a 1,2-acetal and with benzaldehyde to give a mixture of 1,2- and 1,3-acetals which can be separated by crystallisation. These protecting groups can be removed by hydrolysis with aqueous acid though this may promote acyl migration. The benzylidene groups can also be removed by hydrogenolysis, but this is inappropriate when an unsaturated acyl group has been introduced. These difficulties are overcome in the milder reaction with boric acid in trimethyl or triethyl borate to give a borate ester hydrolysable by water at room temperature.

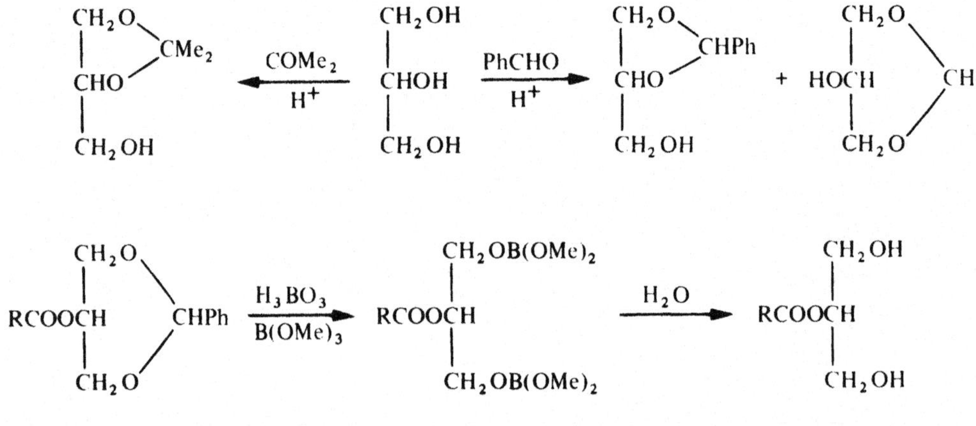

The 1- and 2-benzylglycerols are prepared from 1,2-isopropylideneglycerol and 1,3-benzylideneglycerol respectively by benzylation followed by removal of the protecting group. The benzyl group is removed by hydrogenolysis when it has served its purpose.

Glycerol, triphenylmethyl chloride (trityl chloride), and pyridine furnish 1-trityl and 1,3-ditrityl derivatives when reacted together at room temperature. The trityl group can be removed by reaction with hydrogen bromide in acetic acid or with hydrogen chloride in ether or in petrol, by hydrogenolysis, or by percolation through a column of fresh silica.

Glycerol carbonate, resulting from reaction with phosgene ($COCl_2$), can be decomposed by mild alkaline hydrolysis. The 2,2,2-trichloroethylcarbonyl protecting group is removed with zinc and acetic acid.

As an alternative to the use of protecting groups, glycerol may be replaced by other C_3 compounds in which hydroxyl groups can be generated as required. Examples include 1,3-dihydroxy-propan-2-one (dihydroxy acetone), prop-2-en-1-ol (allyl alcohol), 2,3-epoxypropan-1-ol (glycidol), and 3-chloropropane-1,2-diol.

4. Monoacylglycerols

1-Monoacyl- and 2-monoacylglycerols are usually prepared from 1,2-isopropylideneglycerol and 1,3-benzylideneglycerol as set out in Schemes 10.1 and 10.2.

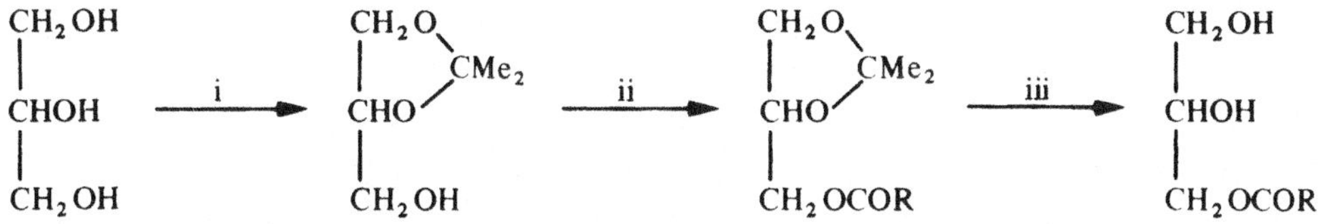

i, $COMe_2$, C_6H_6, p-$MeC_6H_4SO_3H$; ii, RCO_2H, p-$MeC_6H_4SO_3H$; iii, H_3BO_3, $B(OMe)_3$; H_2O.

SCHEME 10.1. Preparation of 1-monoacylglycerols.

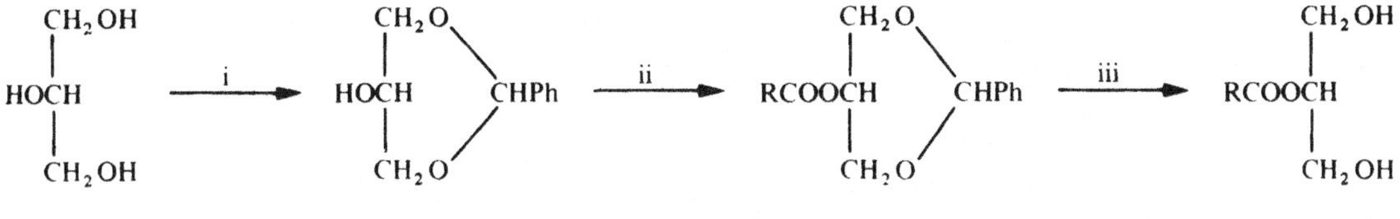

i, PhCHO, p-$MeC_6H_4SO_3H$; ii, RCOCl, pyridine; iii, H_3BO_3, $B(OMe)_3$; H_2O.

SCHEME 10.2. Preparation of 2-monoacylglycerols.

5. 1,3-Diacylglycerols

1,3-Diacylglycerols with two different acyl groups are prepared by direct acylation of the appropriate 1-monoacylglycerol (Scheme 10.3). Diacylglycerols with two identical acyl groups are obtained from dihydroxyacetone or by direct acylation of glycerol (Scheme 10.4). Some of these syntheses depend on the greater reactivity of the primary hydroxyl groups and on the ease with which the 1,3-diacylglycerols can be purified by crystallisation.

i, R^1COCl, pyridine.

SCHEME 10.3. Synthesis of 1,3-diacylglycerols with two different acyl groups.

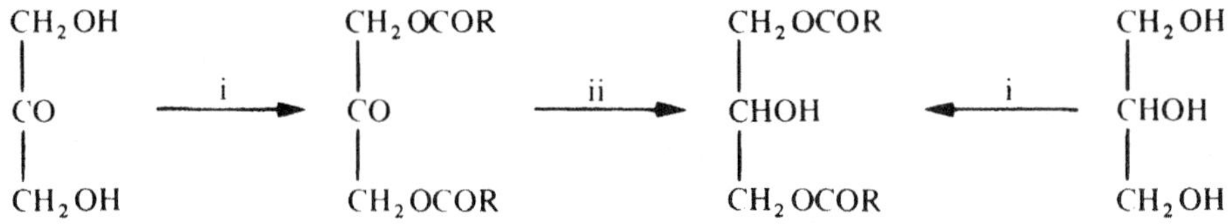

i, RCOCl, pyridine; ii, $NaBH_4$

SCHEME 10.4. Synthesis of 1,3-diacylglycerols with two identical acyl groups.

6. 1,2-Diacylglycerols

These are more difficult to prepare than their 1,3-isomers because of the readiness with which they undergo acyl migration and because they are not so easily crystallised. Four routes are formulated. The first from benzylglycerol is only suitable for compounds with saturated acyl groups. The second starting with allyl alcohol can give a diacylglycerol with two identical groups—saturated or unsaturated—or the acyl group attached to C-1 can be removed by pancreatic lipase hydrolysis and replaced by a different acyl group. Syntheses starting with 2,3-epoxypropan-1-ol(glycidol) or 3-chloropropane-1,3-diol give products with identical or different acyl groups.

7. Triacylglycerols

Monoacid triacylglycerols are easily made by direct acylation of glycerol with acid chloride, anhydride, or free acid. Di- and tri-acid triacylglycerols are obtained by extension of the procedures already described for mono- and diacylglycerols. In designing a synthetic route the aim should be to proceed with the minimum number of steps through the most stable intermediates (avoid 1,2-diacylglycerols) and to introduce unsaturated acyl groups as late as possible. These principles are illustrated in the typical sequences collected in Scheme 10.6.

8. Optically active acylglycerols

The methods discussed so far produce only racemic glycerides and special procedures are required to prepare enantiomeric acylglycerols. Many of these compounds—particularly the triacylglycerols—have a very small optical rotation which is difficult to measure, so their

Route 1

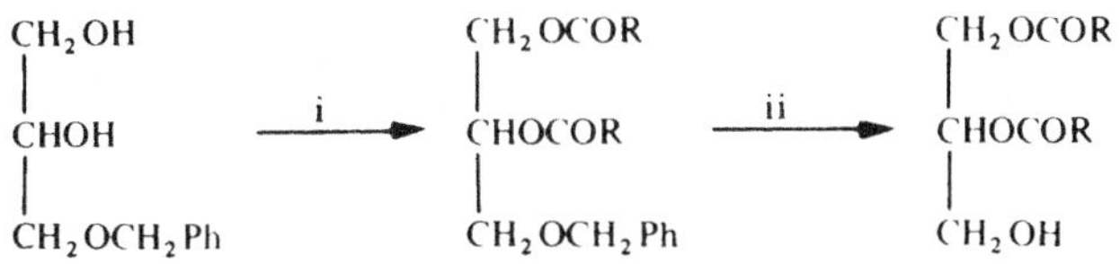

Route 2

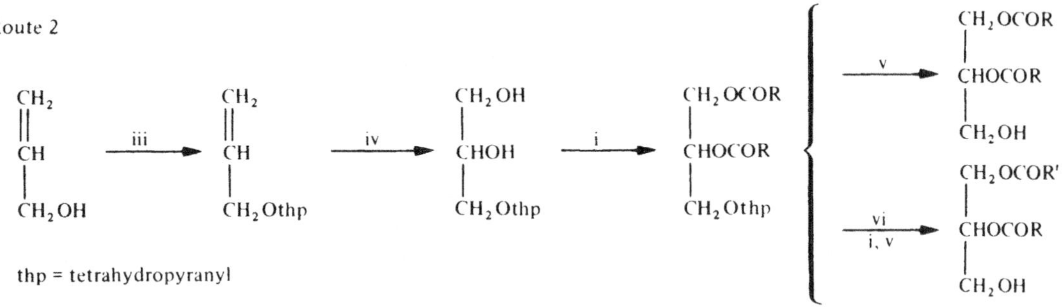

Routes 3 and 4

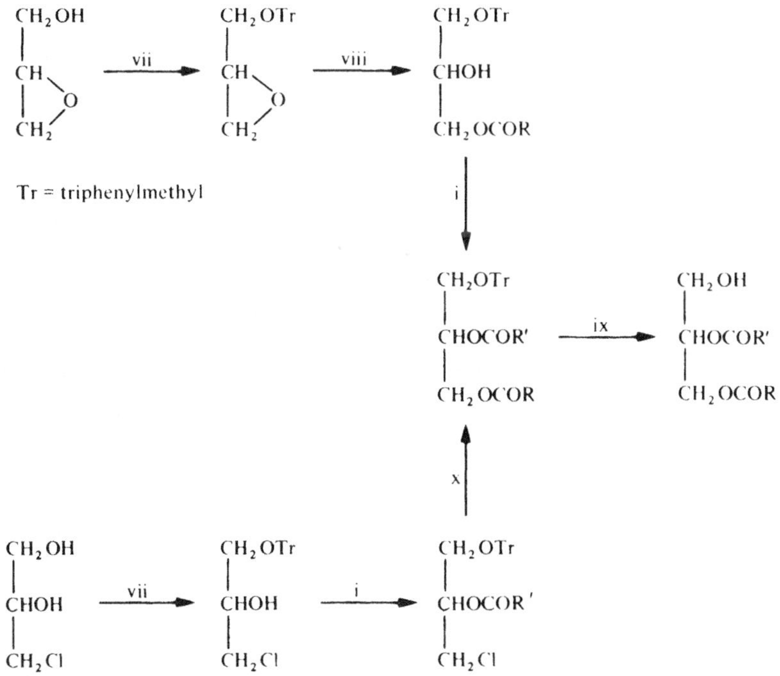

i, acyl halide, pyridine; ii, Ni, H_2; iii, dihydropropan, H^+; iv, $KMnO_4$; v, HCl or $B(OH)_3$; vi, pancreatic lipase; vii, Ph_3CCl; viii, RCOOH; ix, $(CF_3CO)_2O$, MeOH; x, RCOONa

SCHEME 10.5. Preparation of 1,2-diacylglycerols.

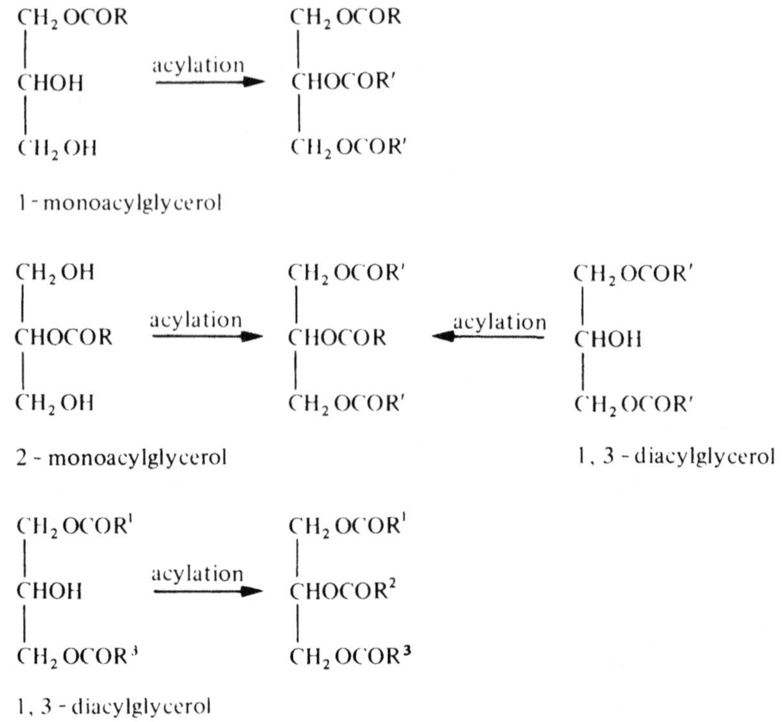

SCHEME 10.6. Preparation of triacylglycerols.

stereochemical integrity must be assessed in other ways such as enzymic hydrolysis or nmr spectroscopy.

The synthesis of enantiomeric acylglycerols has been based on the preparation of 1,2-isopropylidene-*sn*-glycerol from D-(+)-mannitol (Scheme 10.7) and, when necessary, 2,3-isopropylidene-*sn*-glycerol from the less readily available L-(−)-mannitol. These enantiomeric glycerol derivatives furnish enantiomeric acylglycerols by application of the methods already described for racemic compounds. The synthesis of 1,2-diacylglycerols (Scheme 10.8) is of special interest because such compounds are required for the synthesis of enantiomeric phosphoglycerides (Section D).

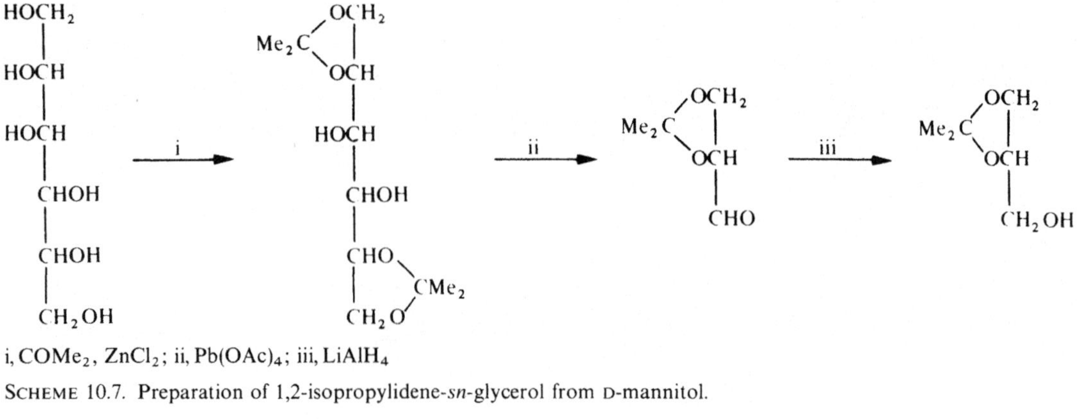

i, $COMe_2$, $ZnCl_2$; ii, $Pb(OAc)_4$; iii, $LiAlH_4$

SCHEME 10.7. Preparation of 1,2-isopropylidene-*sn*-glycerol from D-mannitol.

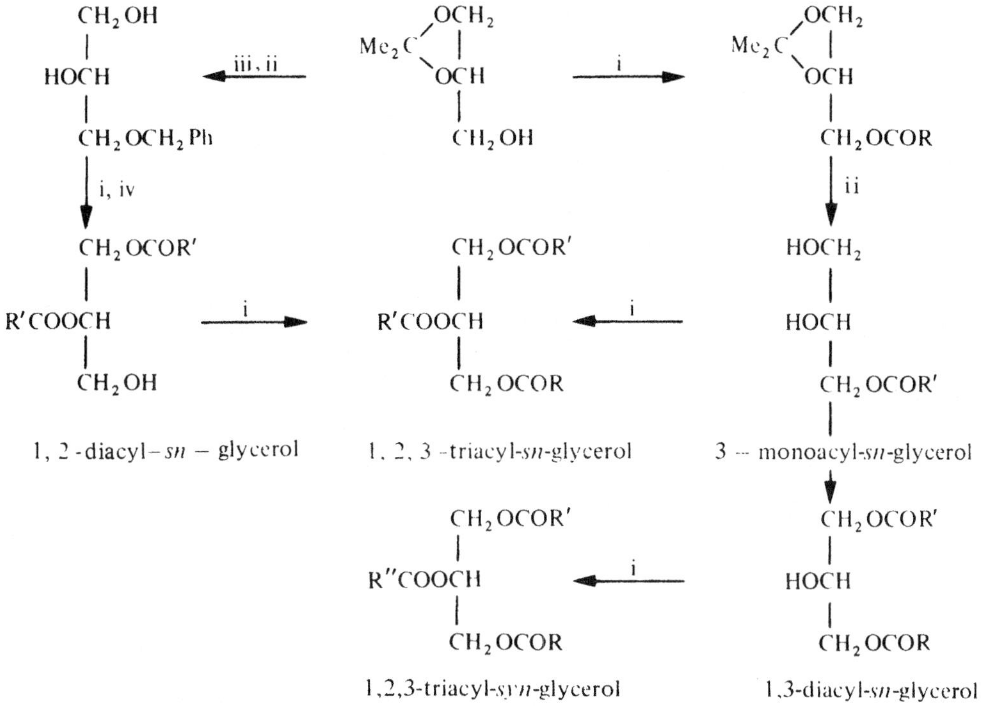

i, acylation; ii, $B(OH)_3$; iii, Na, $PhCH_2OH$; iv, Pd, H_2

SCHEME 10.8. Preparation of mono-, di-, and triacyl-*sn*-glycerols.

A new approach to chiral glycerides (Scheme 10.9), based on the preparation of D- and L-glycidol (2,3-epoxypropan-1-ol) from L- and D-serine (2-amino-3-hydroxypropanoic acid) has the advantage that both isomers are readily available.

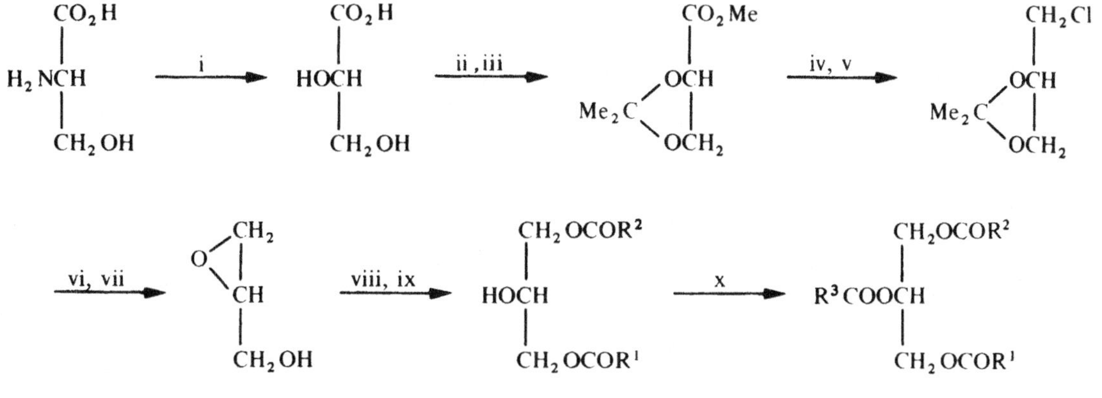

i, $NaNO_2$, HCl; ii, MeOH, $Me_2C(OMe)_2$; iii, $COMe_2$, $Me_2C(OMe)_2$; iv, $LiAlH_4$; v, Ph_3P, CCl_4; vi, aq. AcOH; vii, Na, Et_2O; viii, R^1COCl; ix, R^2CO_2H, $Et_4\overset{+}{N}\overset{-}{Br}$; x, R^3COCl.

SCHEME 10.9. Preparation of 1,3-diacyl- and 1,2,3-triacyl-*sn*-glycerols from L-serine.

D. SYNTHESIS OF PHOSPHOGLYCERIDES[6,7]

1. Introduction

This brief account is restricted to the more important methods of preparing phosphatidyl esters. These compounds have four ester linkages and the major preparative procedures vary in the order in which these are assembled. The most common starting materials are 1,2-diacylglycerols (or a related iodo compound), glycerophosphocholine (ethanolamine), or phosphatidic acids. When the phospho-linked alcohol contains additional functional groups these must be blocked during phosphorylation of the hydroxyl group. Optically active compounds are prepared from enantiomeric starting materials prepared as already indicated or obtained from natural lipids. The simplest way of producing phosphatidyl esters with different acyl groups involves selective enzymic deacylation and chemical reacylation. Products are purified by crystallisation and/or column chromatography and purity is checked by thin-layer chromatography, hydrolysis with phospholipase A, total fatty acid determination, specific rotation, and P/N ratio.

2. Preparation from 1,2-diacylglycerols

Diacylglycerols can be converted to phosphatidyl esters by phosphorylation with phosphorus oxychloride or monophenylphosphoryl chloride followed by reaction with an appropriate hydroxy compound or directly with an appropriate phosphate derivative. Some reactive centres may have to be blocked and the blocking group subsequently removed (Scheme 10.10).

Better results are obtained when the diacylglycerol is converted to its iododeoxy derivative and then treated with a suitable phosphate ester (Scheme 10.11).

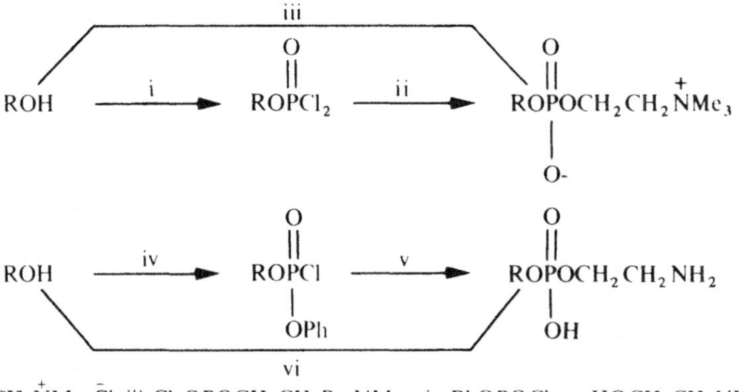

i, $POCl_3$; ii, $HOCH_2CH_2\overset{+}{N}Me_3\overset{-}{Cl}$; iii, $Cl_2OPOCH_2CH_2Br$, NMe_3; iv, $PhOPOCl_2$; v, $HOCH_2CH_2NHCOOCH_2Ph$; H_2, Pd; vi, $Cl_2OPOCH_2CH_2N(CO)_2C_6H_4$; N_2H_4

SCHEME 10.10. Preparation of phosphatidyl esters from 1,2-diacylglycerols (represented as ROH).

3. Acylation of glycerolphosphocholine or glycerolphosphoethanolamine

Pure phosphatidylcholine can be obtained from egg yolk. This is a mixture of compounds consisting entirely of one lipid class. A phosphatidylethanolamine preparation can be obtained

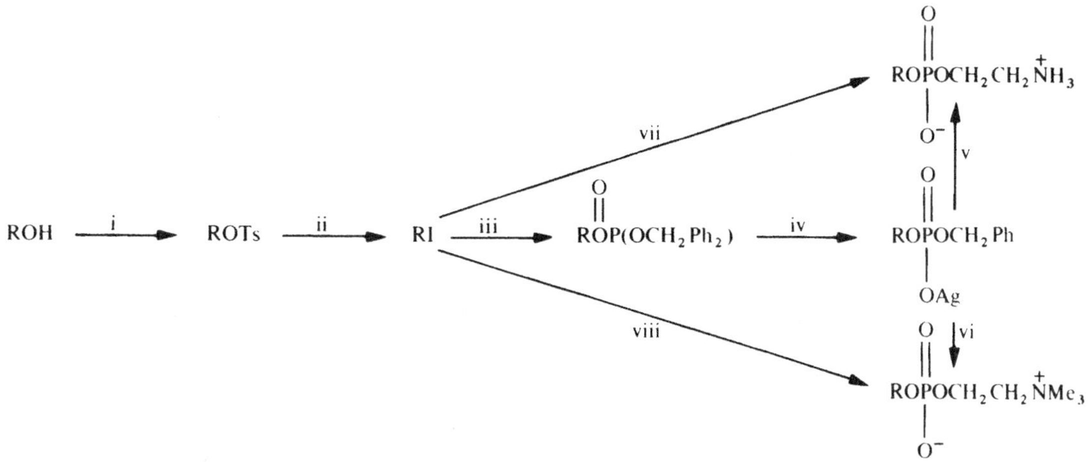

i, TsCl; ii, NaI; iii, $AgOPO(OCH_2Ph)_2$; iv, NaI, $AgNO_3$; v, $BrCH_2CH_2N(CH_2Ph)_2$; H_2/Pd; vi, $BrCH_2CH_2\overset{+}{N}Me_3$picrate$^-$; NaI; vii, $AgOP(OY)CH_2CH_2NHX$ (Y = Bu^t or Ph or $PhCH_2$, X = $(CO)_2C_6H_4$ or $COOCH_2Ph$ or CPh_3); viii, $AgOP(O)\ (OCH_2C_6H_4NO_2{-}4)OCH_2CH_2Cl$; NMe_3

SCHEME 10.11. Preparation of phosphatidyl esters from 1,2-diacyloxy-3-iodopropane (represented as RI).

from the same source with a little more difficulty. The natural phospholipids can then be deacylated with tetrabutylammonium hydroxide to give glycerolphosphocholine and glycerolphosphoethanolamine in enantiomeric form (the *sn*-3 isomers). These compounds can be reacylated by heating with a fatty acid anhydride and its potassium salt. If two different acyl groups are required, then selective enzymic degradation is followed by chemical reacylation.

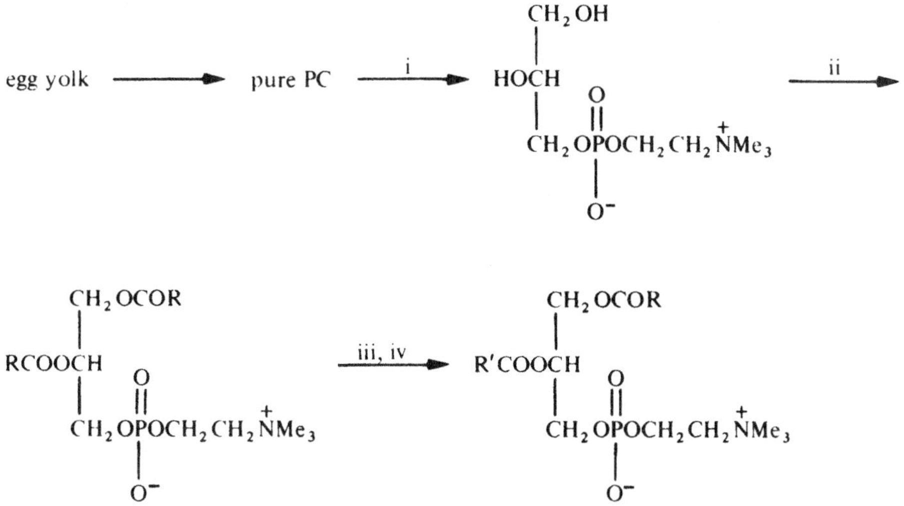

i, $Bu_4\overset{+}{N}\overset{-}{O}H$; ii, $(RCO)_2O$,RCOOK; iii, phospholipase A_2; iv, $(R'CO)_2O$,R'COOK

SCHEME 10.12. Preparation of phosphatidylcholines from glycerophosphocholine.

4. Preparation of phosphatidic acids and of phosphatidyl esters therefrom

Phosphatidic acids are prepared from 1,2-diacylglycerols or their iododeoxy derivatives as shown in Scheme 10.13. These can be converted to phosphatidyl esters by reaction with appropriate reagents.

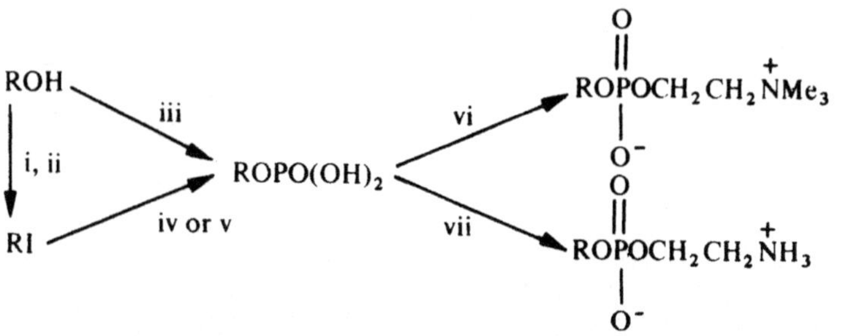

i, TsCl; ii, NaI; iii, $POCl_3$; H_2O; iv, $AgOPO(OCH_2Ph)_2$; BaI_2; v, $AgOPO(OBu^t)_2$; HCl; vi, 2,4,6-$(Pr^i)_3C_6H_2SO_2Cl$, $HOCH_2CH_2\overset{+}{N}Me_3\overset{-}{O}Ac$; vii, 2,4,6-$(Pr^i)_3C_6H_2SO_2Cl$, $HOCH_2CH_2NHCPh_3$; H_2/Pd.

SCHEME 10.13. Preparation of phosphatidic acids and esters (ROH represents a 1,2-diacylglycerol).

REFERENCES

1. H. SPRECHER, *Progress in the Chemistry of Fats and Other Lipids*, 1979, **15**, 219.
2. W. J. DE JARLAIS, *Fatty Acids* (ed. E. H. Pryde), American Oil Chemists' Society, Champaign, 1979, p. 76.
3. E. A. EMKEN, *Fatty Acids* (ed. E. H. Pryde), American Oil Chemists' Society, Champaign, 1979, p. 90.
4. R. G. JENSEN, *Topics Lipid Chem.* 1972, **3**, 1.
5. R. G. JENSEN and R. E. PITAS, *Adv. Lipid Res.* 1976, **14**, 213.
6. R. G. JENSEN and D. T. GORDON, *Lipids*, 1972, **7**, 611.
7. A. J. SLOTBOOM, H. M. VERHEIJ, and G. H. de HAAS, *Chem. Phys. Lipids*, 1973, **11**, 295.

CHAPTER 11

Recovery of Fats and Oils from Their Sources

A. INTRODUCTION

The lipids that make up the fat portion of human diets are derived from dairy products, tropical plants, oilseeds, animals, and fish. While some, like peanuts, can be eaten in their natural state, admixed with proteins and carbohydrates, most are processed to separate the fat portion. This is then commonly further purified to remove components that adversely affect colour, flavour, nutritional value, stability, or ease of handling, including incorporation into other food products. The technology employed varies considerably with the source material, its geographical location, monetary value, and considerations of by-products from the processing. For example, cottonseed is grown for the fibre and not the oil; hence, oil recovery must not hurt fibre yield and quality. In addition, the cottonseed meal produced as a by-product must be of sufficient nutritional value to be readily saleable without penalty. Similarly, when soybeans are processed for oil, roughly five times as much meal as oil is produced. If the meal were not marketable, the resultant oil price would not be competitive with other similar oils and the processing would probably be impractical.

Oils and fats occur in oilseeds (18–70 %), fruit pulp (30–58 %), animal tissues (60–90 %), and fish (10–20 %). The method for removing the oil is largely dependent upon the source. Fatty animal tissues consist largely of fat and water which may be separated from the solid portions of the tissue and from each other rather easily by "rendering" (dry or wet heat treatment). Oilseeds tend to contain a much larger proportion of solid material associated with the oil, requiring careful reduction in size and usually some heat treatment before being pressed or solvent extracted to recover the oil.

In order of world consumption, the relative amounts of fats and oils consumed are about as shown in Table 11.1.

Per capita consumption of edible fats and oils varies from 21 lb per year in Japan up to 68 lb per year in The Netherlands. The United Kingdom and the United States consumptions are 56.3 lb and 53.7 lb, respectively.[2]

B. METHODS OF OBTAINING CRUDE FATS AND OILS

1. General

Oilseeds contain enzymes (lipases) which, in the presence of moisture, can break down glycerides liberating free fatty acids (FFA). Since only intact triglycerides are normally desired for food, any

TABLE 11.1. *Fats and Oils Consumption—1980*[1]*†

Oil or fat	World consumption	
	Billions of pounds	% of total world
Soybean	28.04	23.7
Tallow (incl. inedible)‡	13.53	11.4
Butter	12.64	10.7
Sunflower	11.08	9.4
Lard	10.00	8.5
Rapeseed	7.72	6.5
Palm	7.05	6.0
Cottonseed	6.98	5.9
Coconut	6.03	5.1
Peanut	5.84	4.9
Olive	3.89	3.3
Fish	2.48	2.1
Sesame	1.23	1.0
Palm kernel	1.23	1.0
Corn ∫	0.60	0.5
	118.34	100.0

* Superscript numbers refer to References at end of Chapter.
† Quantities are taken from statistical tables on "disappearance", assuming that, except for inedible oils and fats, "disappearance" is roughly equivalent to "consumption". In general, "disappearance" represents the sum of beginning stocks on hand plus production during the year, minus year end stocks.
‡ Edible and inedible disappearance are not given separately.
∫ Estimate by author based on United States production.

breakdown is unwanted and expensive. Hence, oilseeds are normally dried to a 7–13% moisture range for good storage and processing. The aim for oil mills is to operate them year round, often necessitating rather prolonged storage in the plant or prolonged travel time in the case of imports. Also, oilseeds like cottonseed can "heat" as the result of enzyme action, even to the point of igniting and partial or complete loss by fire. Less extensive heating can adversely affect oil quality in addition to the breakdown of the glycerides.

Oilseeds as received at the mill are normally admixed with materials accumulated during harvesting and transportation. They must be cleaned of tramp (trash) metal, sticks, stones, and weed seeds (dockage) using magnets, screens, and aspirator systems. Also, as in the case of soybeans, for example, there is an outside seed coat ("hull") which is high in fibre. If this is not removed before the oil is extracted, the resulting meal is relatively lower in protein and higher in fibre than common Trading Rules allow (minimum 44% protein and maximum 7% fibre), limiting its feed utilisation. On the other hand, if the soybeans are dehulled, higher protein meal can be produced (49% minimum) and fibre is reduced (maximum of 3.3%). Trading Rules prescribe the exact premiums and penalties in the United States.[3]

Table 11.2 shows the oil content of some common oil-source materials.

2. Soybean and cottonseed preparation

Preparation has to be tailored to each oilseed, but there are some general guidelines common to most. For illustrative purposes, soybean and cottonseed processing will be discussed here.

TABLE 11.2. *Oil Content of Some Common Oil Sources*

	%
Babassu	60–65
Copra	65–68
Corn	5
Cottonseed	18–20
Olive	25–30
Palm, fruit	45–50
Palm, kernel	45–50
Peanut	45–50
Rapeseed	40–45
Safflower	30–35
Sesame	50–55
Soybean	18–20
Sunflower	35–45

Soybeans are usually stored in large concrete silos until they are processed. Then they are normally dried to about 10% moisture and tempered several days to facilitate dehulling, which is commonly done before oil removal. Dehulling increases capacity by removing this bulky and essentially oil-free material from the feed, while also producing a more desirable low fibre, higher oil-content meal. Dehulling is accomplished by feeding the dried beans to cracking rolls where the soybeans are broken into quarters and eighths. The cracked beans move to a shaker screen where the fine particles drop out and the hulls rise to the top where they are separated by aspiration. Several systems are in use to make a clean separation of hulls without losing any appreciable quantity of the high oil-content meats.

It must also be mentioned that hulls can be separated by aspiration after solvent extraction of the oil ("tail end dehulling"), but this is useful mainly where there is a low demand for high protein–low fibre meal. By this method only enough meal is dehulled to satisfy demand and the plant does not need to dehull completely and later add back hulls to the meal in accordance with trade preference.

The cracked beans are conditioned by heating, commonly in steam-jacketed vertical stack-type cookers or rotary steam tube dryers (conditioners) for 20–30 min, reaching a temperature of 70–75°C (158–167°F). Discharge moisture is about 11%. These warm beans are easier to flake, being more plastic and less susceptible to breaking down into "fines". Flaking is accomplished in roll stands, each consisting of a pair of smooth-surface rolls regulated to produce flakes about 0.254 mm (0.010 in) thick in one pass. Optimum flake thickness varies with the equipment used for oil extraction. Large volume deep bed extractors, such as the Rotocel and the French stationary basket extractors, can use thicker flakes (to 0.4 mm (0.0157 in)), since these thick and dense flakes pack into the extractor, permitting a longer extraction time which compensates for the thicker flakes. Smaller capacity plants usually need a flake thickness of 0.25–0.35 mm (0.010–0.014 in) for efficient extraction. After leaving the rolls the flakes are carried to the extraction unit in mass flow-type enclosed conveyors to minimise breakage. A flowsheet of soybean and cottonseed processing is shown in Fig. 11.1.

When cottonseed is processed, it must first be delinted, i.e. freed of most of the cellulose fibre covering the oil-containing meats. This is done using several stands of "linters", each consisting essentially of a revolving assembly of closely-spaced circular saws which pick the lint from the seed. The fibres are removed from the saw teeth by aspiration. Normally, lint is removed in two or three cuts of decreasing fibre length, each successive cut being of lower grade and price than the preceding one. The delinted seeds are then partially dehulled, leaving meats containing 10–15%

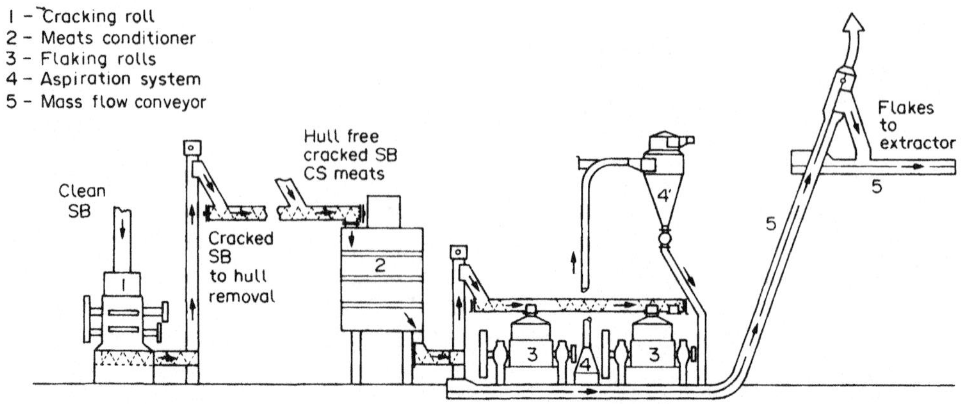

FIG. 11.1. Flow chart of soybean and cottonseed meats preparation for direct extraction. (Courtesy of the French Oil Mill Machinery Co., Piqua, Ohio.)

hulls and about 33 % oil. These meats are then conditioned at a maximum temperature of 65°C (149°F) for 10–12 min. Moisture should preferably be in the 8–11 % range. The meats are flaked to about 0.2 mm (0.008 in) thickness for solvent extraction. These flakes are considerably more fragile than soybean flakes and must be handled carefully in mass flow-type conveyers.

Other oilseeds are prepared for oil removal along generally similar lines, although details vary with the type, oil content, hull content, size, etc.

3. Recovery of oil from oilseeds

(a) *Mechanical methods.* The oldest method of oil extraction was the application of pressure to batches of oil-bearing material confined in bags, cloths, cages, etc. Levers, wedges, screws, etc., were used as a means of applying the pressure. Gradually, the hydraulic ram system was adopted, using pressures from 500 to 6000 psi or more; however, this was a batch operation with resultant high labour costs. In addition, up to about 5–10 % oil was left in the press cake and drainage time had to be carefully controlled for consistent results. Today, hydraulic processing is little used and mainly for low volume specialty materials.

The first continuous mechanical press called an "expeller" was manufactured by V. D. Anderson in the United States in 1900 and was used to express oil from flaxseed and whole cottonseed. In 1910 Krupp licensed these machines in Germany where they were used primarily as a forepress unit ahead of hydraulic presses. Many improvements were made as time passed, and in 1933 a new United States competitor, the French Oil Mill Machinery Company, introduced the "screw press". Since that time many improvements in efficiency and volume have been made by manufacturers throughout the world. An expeller or "screw press" is essentially a continuous device for gradually increasing the pressure on material fed to it as the latter progresses inside a closed barrel, with provisions for the oil to drain out as it is squeezed from the feedstock. Depending on the use, machines can reduce oil content to the 3 % range and capacities may be as high as 460 short tons per day when used for pre-pressing, where oil content is reduced to only 10–15 % in preparation for solvent extraction.

Figure 11.2 shows a side view of an uncovered mechanical screw press. Feed enters through a vertical feed screw shown on the right and is force fed to the revolving horizontal worm shaft which

FIG. 11.2. Side view of a D-88 mechanical screw press for full pressing showing drainage bars and water cooling. (Courtesy of the French Oil Mill Machinery Co., Piqua, Ohio.)

moves it down the barrel of the cage by the pushing action of the flights on the worms of the shaft arrangement (Fig. 11.3). The cage is water-cooled by attachments as shown. The oil extracted from the material is drained through the slots in the cage. The cake discharges out of the cone located on the left end of the machine.

In full continuous pressing the oil-bearing material is usually flaked and then "cooked" with a high moisture content for about 20 min at a temperature of about 87°C (189°F). This inhibits enzyme action and also coagulates the proteins so that the oil may be squeezed out in reasonably pure form. Heating is then continued, usually in a stacked steam jacketed "cooker" until the moisture drops to around 3.0%. The temperature will reach 115°C (240°F). The material, now ready for pressing, is conveyed to a cage area where pressure is gradually built up to a maximum (15,000–20,000 psi) by continuously rotating worm shafts and worms, plus a choke mechanism by means of which the thickness of the exit cake is controlled. The processing cages may be cooled with water or the expressed oil. Variations in the worm arrangements, shaft length, and bar spacings depend upon the oilseed processed, rate, and whether full or pre-pressing is desired.

While better than hydraulic pressing, continuous mechanical pressing is not as efficient as solvent extraction, but it usually has the advantages of lower installed cost, less safety hazards, is easily adaptable for small capacities and changes in capacity, and generally requires less skilled labour. On the other hand, oil yield is lower, especially for lower oil content seeds, since the residual oil content is usually in the 3% range, versus 1% for solvent extraction. The current trend is to use high capacity continuous pressing to squeeze the easily available oil out of high oil content oilseeds, leaving a press cake that does not tend to break down into "fines" and from which the remaining oil (10–15%) is easily extractable with a somewhat smaller solvent plant. For example, in the United States cottonseed is usually pre-pressed before extraction rather than extracted directly.

Screw presses currently have capacities of 22–110 tons per day for full pressing and 100–460 tons per day when used as pre-presses. Figure 11.4 shows the worm arrangement on a B-2100 screw press used for pre-pressing. Capacity is 170 tons per day. Feed enters from the hopper on the right and is compacted as it moves along the shaft of slowly increasing diameter. Expressed oil leaves through the slots in the cage and cake leaves through a cone mechanism on the left where the cake is discharged.

(b) *Solvent extraction.* (i) *Selection of solvent.* The analytical determination of oil content is done by extracting a finely ground dry sample with a boiling fat solvent, such as petroleum ether. Commercial solvent extraction is similar except that, for practical purposes, the material is not ground as fine or dried as much, the solvent is not brought to boiling, and by various contact procedures proportionately less solvent is required. In addition, a solvent is selected which can be recovered with available condenser water and which is not so high boiling as to be difficult to strip from the oil and extracted residue. The most common solvent is hexane, boiling point 63–69°C (146–156°F). Where condenser water is too warm, heptane (boiling point 90–99°C or 194–210°F) is sometimes used. Many other fat solvents are either too flammable or not selective enough in what they extract. A non-flammable solvent would be most welcome and trichlorethylene (boiling point 86.7°C (188°F)) has been used in Europe and in the United States. This has given rise, however, to toxicity problems when soybeans are extracted, as well as causing corrosion problems. Its use is now prohibited in the United States.

(ii) *Extraction principles.* Extractor design is determined by the rate at which equilibrium is attained between a lean miscella (oil–solvent mixture) outside the seed particles and the solvent and oil within the particles. As the oil content is reduced to commercially acceptable levels (usually 1% or below), equilibrium is obtained slowly, in part due to the relative insolubility of the last remaining "oil". Actually, soybean "oil" obtained from beans previously extracted down to 0.5 oil is a solid

FIG. 11.3. Opened cage half on a D-88 screw press showing the arrangement of flights on the shaft to compress the oil-containing feed increasingly as it is moved to the exit (cake) end. (Courtesy of the French Oil Mill Machinery Co., Piqua, Ohio.)

FIG. 11.4. Opened cage half on a B-2100 screw press used for pre-pressing. (Courtesy of the French Oil Mill Machinery Co., Piqua, Ohio.)

containing much non-triglyceride material and it is not economical to add this to the other oil.

The rate at which equilibrium is approached is affected by the intrinsic capacity for diffusion of solvent and oil, the size and shape of the seed particles, their internal structure, and, finally, the solubility of the last remaining oil. Despite many excellent theoretical studies, extractor design must consider practical conditions in available equipment. For example, it has been claimed that extraction time varies as the square of the flake thickness. Thus, doubling the flake thickness quadruples the extraction time. On this basis, flakes should be as thin as possible; however, at levels much below 0.25 mm (0.010 in), flakes are usually so fragile that they break up into "fines" in the equipment, drain poorly, and cause major miscella filtration problems. On the other hand, so-called "thick" flakes (to 0.4 mm, or 0.0157 in) pack better in deep bed extractors, allowing longer contact time which counterbalances the adverse effect of the thicker flakes.

It is generally agreed that hot solvent is better than cold. It has also been claimed that extraction time varies inversely as the square of the extraction temperature in degrees Fahrenheit. Thus, if 40 min are required at 145°F, going to 120°F would increase extraction time to over 58 min. Generally, temperatures near the boiling point are preferred.

Conventional countercurrent extraction is based on the theory that the less oil in the miscella contacting flakes, the faster the oil is removed from the flakes. On the other hand, it has been claimed[4] that the concentration of oil in the miscella has no effect on either total extraction or extraction rate. Thus, if there is an intimate mixture of solids and solvent for enough time, the retained miscella on the surface of the flakes should be removable by washing, leaving low oil content extracted flakes. This has yet to be properly evaluated in commercial size extraction, but deserves more study.

(iii) *Types of extractors.* Two general types of extractors exist:

(a) Percolation extractors. This is the most common design used today. Liquid solvent or miscella is pumped over a bed of flakes or cake and percolates down through the bed leacing at the bottom through a perforated plate or some type of screen system. Variations of this type include chain and basket, perforated belt, rotary, chain conveyer, and filter types. The percolation system is generally considered to be more adaptable to large capacities in limited space. Examples are the French, Dravo, De Smet, Crown, and Wurster and Sanger units.

(b) Immersion extractors. In this type the material to be extracted is moved through a pool of extracting solvent, usually by chain or screw conveyers. It is sometimes used in connection with a percolation extractor, as in the Bernardini type. An example is the EMI horizontal extractor used for a variety of special extractions. In general, immersion extractors are less adaptable to oilseed like cottonseed where flakes are extremely fragile, breaking up into many "fines" which must be removed from the miscella before recovery of the oil.

(iv) *Typical procedure.* In a common type of percolation extractor, soybean flakes entering the extraction area are dropped into a filling screw feeder which is designed to provide a plug seal of material (Fig. 11.5). The flakes are then slurried into the basket compartments of the extractor, through a filling spout, where they are extracted by the application of miscellas in diminishing strengths and finally with a wash of fresh solvent to reduce the residual oil to less than 1 %. Following a drainage period to reduce the solvent retention of the flakes to 35–40 %, the flakes are discharged into a vapour-tight mass-flow type conveyor with a plug seal, which feeds the meal desolventising apparatus. The latter is commonly a desolventiser-toaster, consisting of a vertical stacked cooker with agitation in each section. In the upper stages live steam is used to strip off the hexane, condensing as it gives off its heat, thus increasing the moisture content of the flakes to the 25 % range. In the middle and lower sections, indirect heat toasts and dries the flakes to an exit moisture level of 10–12 %. This treatment inactivates antigrowth factors present and improves the palatability of the meal for animals; however, the toasting procedure, while increasing relative

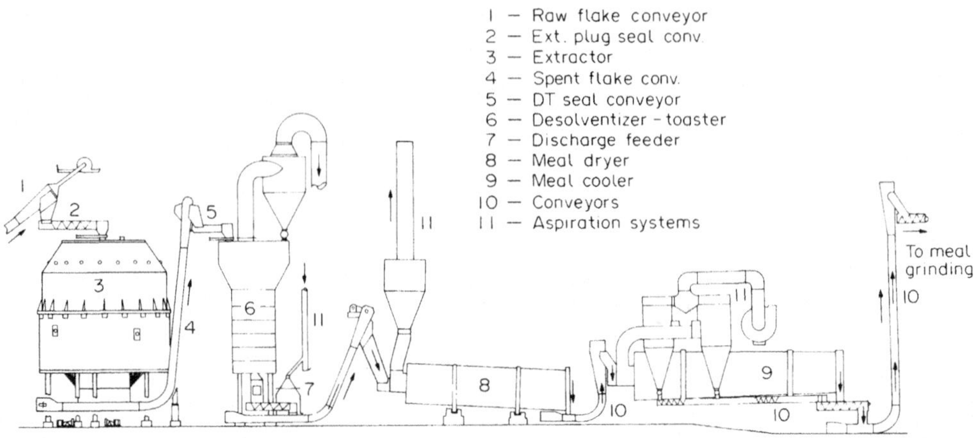

FIG. 11.5. Extraction solids flow typical for large capacity soybean operation. (Courtesy of the French Oil Mill Machinery Co., Piqua, Ohio.)

protein efficiency from the 40–50 % to the 85–90 % range, also denatures the protein, lowering the protein dispersibility index (PDI) from the original 90–95 % range to 8–20 %.[5] Where more functionality is desired, through a higher PDI, the meal may be desolventised by superheated solvent vapour or by a flash desolventising to produce a final meal of almost any PDI up to that within 1–2 % of the original extracted flakes.

The final miscella (solvent–oil mixture) (Figure 11.6) leaving the extractor may contain 20–35 % oil as well as a variable amount of fine meal particles. These are normally removed by a liquid cyclone or filter. The clear miscella is typically concentrated to about 90 % in two stages using indirect heat and some vacuum. It is then pumped to a disc and doughnut oil stripping column operated at about one-third atmospheric pressure where the remaining solvent is removed and the moisture and volatile matter (M & V) is reduced to about 0.3 %. Crude soybean oil leaving the stripper need not be cooled immediately, but some other oils deteriorate in quality if not cooled quickly. Cottonseed, for example, will darken to colours difficult to remove by subsequent refining and bleaching.

A flow chart showing pre-press and direct solvent extraction is shown in Fig. 11.7.

Commercial solvent extraction equipment is sold by many firms too numerous to list, but a few are Dravo Corporation, French Oil Mill Machinery Company, Extraktionstechnick, EMI Corporation, Krupp, Lurgi, and Simon-Rosedowns. Plants range in capacity up to about 3000 tons per day.

4. Recovery of oil from fruit pulps

The only fruit pulp oils of commercial importance are olive and palm. Olive oil is produced by milling washed olives into a paste which is pressed to yield a press cake and an oil–water mixture from which the oil is separated. "Virgin" olive oil is from the first pressings. The cake may be extracted with hexane to recover additional oil. Because the fruit is often bruised during handling, olive oil usually has a higher content of free fatty acids than oils such as soybean and cottonseed.

Palm oil, once produced largely by African natives using primitive methods, has undergone major processing changes in line with recent technology so that an oil is produced which is much superior in quality to the older African oil. Thus, plants in Malaysia, Indonesia, and the Republics of the Congo have modernised as increased demand and quality standards have justified.

To prevent free fatty acid build-up, palm fruit is sterilised as soon as possible upon receipt at the processing plant. The fruits and the stalks are then separated using a "picker". Next comes washing, followed by mashing to separate the fruit from the nut and liberate oil from the oil-bearing fruit. This oil is recovered by means of pressing and centrifugation. The oil cakes are freed of fibres and the nuts are dried and crushed to free the kernel which will contain around 50% palm kernel oil. The kernels are dried before extraction.

5. Rendering of animal fats

Unlike oilseeds, fatty animal tissues, free of muscle or bone, are made up of fat (60–90%) plus water and a small amount of connective tissue, largely protein in nature. Tallow is obtained from cattle, and lard and rendered pork fat from pigs. The method used for fat recovery, rendering, varies with the nature of the fat stock and the characteristics wanted in the rendered fat. In "dry" rendering the fatty material, heat alone is used to dry the material and liberate the fat. A household example is the frying of bacon. In "wet" rendering, hot water or steam is used to liberate the fat which is then separated by skimming or a variety of centrifuge methods.

Dry rendering is commonly carried out in horizontal steam-jacketed tanks equipped with an agitator. Liquid fat is drained off and the residue is pressed to yield additional fat. What is left is a protein concentrate ("cracklings") which is ground and used for animal and poultry feed. Dry rendering is used mostly for inedible products where flavour and odour are secondary and where the production of large amounts of high-quality residue is important. Tallow, however, made in

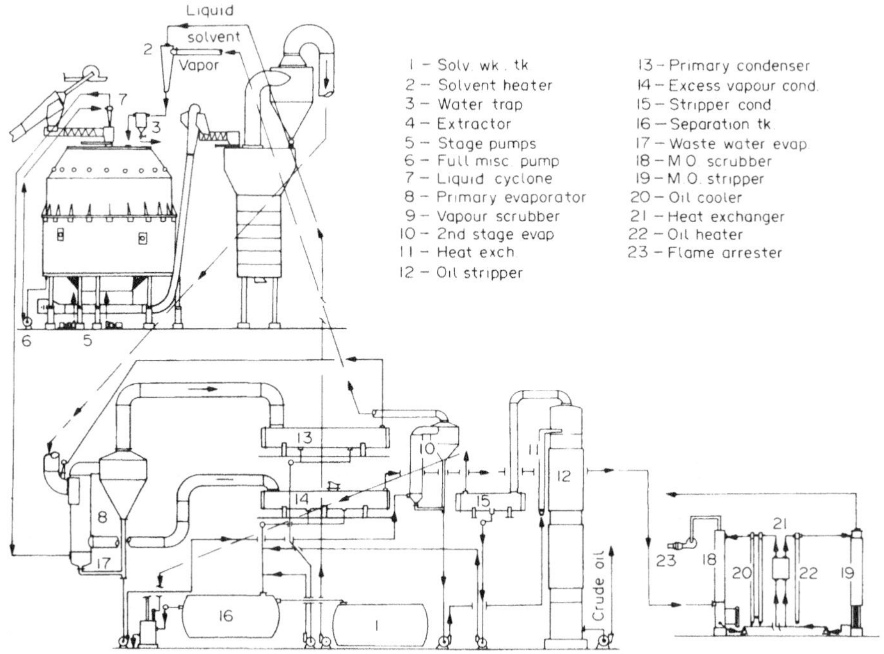

FIG. 11.6. Solvent, miscella and vapour flows during solvent extraction. (Courtesy of the French Oil Mill Machinery Co., Piqua, Ohio.)

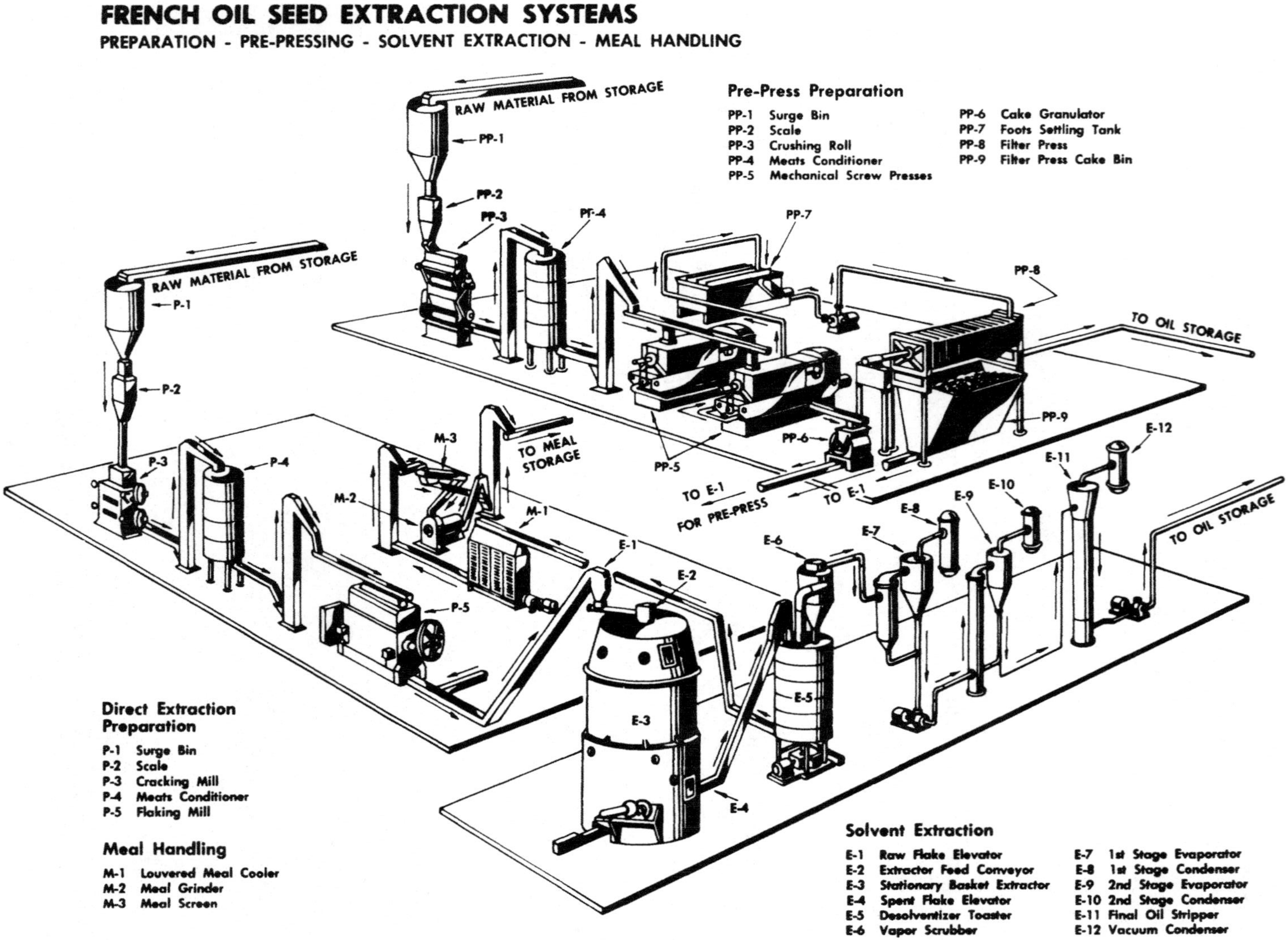

FIG. 11.7. Flow chart of pre-press and direct extraction systems. (Courtesy of French Oil Mill Machinery Co., Piqua, Ohio.)

this way is used extensively in the United States as a frying fat where the flavour imparted to the material fried is quite acceptable, if not preferred.

Wet rendering, using steam, is the method used to produce "prime steam lard", a high-quality edible fat. It is also used extensively for whale oil and tallow. In this process, the fatty material is placed in an autoclave or digester where it is heated under pressure with live steam and a small amount of water. After a digestion period of several hours, the fat rises to the top, leaving a layer of solids (tankage) and "stick water" in the bottom. The layers are separated by various means. There are several modifications of this process, each possessing certain cost or quality advantages.

REFERENCES

1. Data for 1980 from *Oil World*, Oil World Publication, Hamburg, Federal Republic of Germany, 1981.
2. K. F. Gander, *J. Amer. Oil Chemists' Soc.* 1976, **53,** 417.
3. National Soybean Processors Association, 1978/1979, *Yearbook and Trading Rules*, 1800 M Street, N.W., Washington, D.C. 20036.
4. D. F. Othmer and J. C. Agarwal, *Chem. Eng. Prog.* 1955, **51,** 372.
5. G. C. Mustakas, *J. Amer. Oil Chemists' Soc.* 1971, **48,** 815.

GENERAL REFERENCES

Bailey, A. E. (ed.) *Cottonseed and Cottonseed Products*, Interscience, New York, 1948.

Erickson, D. R., E. H. Pryde, O. L. Brekke, T. L. Mounts, and R. A. Falb (eds.), *Handbook of Soy Oil Processing and Utilization*, American Soybean Association, St. Louis, and American Oil Chemists' Society, Champaign, 1980.

Swern, Daniel (ed.) *Bailey's Industrial Oil and Fat Products*, 2 volumes, John Wiley & Sons, New York, Chichester, Brisbane, Toronto, 1979, 1982.

Weiss, T. J., *Food Oils and Their Uses*, The Avi Publishing Co., Inc., Westport, Conn., 1970.

World Conference on Oilseed and Vegetable Oil Processing Technology, *J. Amer. Oil Chemists' Soc.* 1976, **53,** 221–461.

CHAPTER 12

Refining

A. INTRODUCTION

Crude fats and oils contain variable amounts of non-glyceride impurities. Some, such as sterols, are relatively inert; others, like tocopherols, are generally desirable; but some, like free fatty acids, phosphatides,* and certain pigments, are objectionable, tending to make the fat or oil dark coloured, susceptible to foaming and smoking on heating, and liable to precipitation of solid material when the oil is heated during processing operations. The object of refining is to remove the objectionable impurities with minimum damage to the neutral oil (glycerides) and tocopherols and minimum loss of oil. In Europe this is called "neutralisation", and "refining" is the term used to include all processing through deodorisation. In the United States the term "refining" generally is limited to treatment for neutralisation. "Bleaching" and "deodorisation" are terms used to describe processing after the neutralisation step.

In alkali refining, the crude oil is mixed with enough alkali to neutralise the fatty acids present and usually to provide some excess. This converts essentially all the free fatty acids to oil-insoluble soaps which can then be separated. In addition, some other acidic substances combine with the alkali and some other impurities are removed by adsorption on the soap formed in the operation.

Phosphatides (gums) are removed by the alkali treatment along with the free fatty acids; however, if desired, most of the phosphatides can be removed by a pre-treatment with water ("degumming"). Under appropriate conditions, phosphatides absorb water, becoming hydrated and oil-insoluble. They can then be separated by centrifugation. The wet gums may be added to meal to increase fat content and minimise dustiness, or they can be used immediately for other products. They must be carefully dried to low moisture levels (1 %) if they are to be stored or sold. The phosphatides from soybean oil constitute what is known to the trade as "lecithin", although the product is a mixture of phospholipids and occluded oil (35 %) and contains only about 16 % phosphatidylcholine (lecithin). The remainder is approximately cephalin (phosphatidylethanolamine) (14 %), phosphatidylinositol (10 %), phytoglycolipids and other minor phosphatides and constituents (17 %), and carbohydrates (7 %), plus some moisture (1 %).

In the United States degumming is usually carried out at the solvent extraction plant if it is done at all. This allows the plant to sell "degummed oil", preferred for export since it does not tend to precipitate any solids (gums) during shipment. The refineries, on the other hand, customarily remove the gums with the free fatty acids during the alkali treatment. Outside the United States, degumming is often done as a separate step in refining, for lower overall oil loss. United States refineries buy crude and degummed domestic soybean oil as the market dictates, but normally mix

* The term phosphatides is used in this chapter to cover all the phospholipids present.

the two prior to alkali refining, with good results. Initial crude oil quality influences the refiners' decision whether or not to degum as a separate step.

B. ALKALI REFINING METHOD

Alkali refining is the most common refining method. It involves mixing the crude or degummed oil with a known quantity and concentration of sodium hydroxide at a definite temperature, for a definite time, and with prescribed agitation conditions. The insoluble soapstock is separated and the oil washed with hot water to remove residual soap. After that the oil is dried to a moisture and volatile (M & V) content of about 0.10%. The alkali treatment is designed to remove the undesirable crude oil impurities without saponifying any neutral oil which would increase refining loss. Laboratory refining may serve as a guide, but plant experience normally dictates the best operating conditions.

The minimum amount of alkali required for neutralisation can be calculated from the free fatty acid (FFA) of the oil to be refined, using the formula:

$$\%\,NaOH = \%\,FFA \times 0.142,$$

where the FFA is expressed in terms of oleic acid (M.W. = 282). For any desired excess of NaOH, the calculation is:

$$\%\,NaOH = \%\,FFA \times 0.142 + \%\,\text{excess NaOH}.$$

Alkali concentration is usually expressed in terms of the Baumé scale (Table 12.1) rather than percent. For soybean oil 12–17° Baumé alkali is used with about 0.12% excess when refining by continuous methods; however, the alkali strength varies widely with the oil processed, its free fatty acid, and the contact time. Very short contact allows greater concentrations and excess as a rule.

TABLE 12.1. *Alkali Concentrations in % NaOH and in Degrees Baumé*

Degrees Baumé at 15°C	% NaOH
10	6.57
12	8.00
14	9.50
16	11.06
18	12.68
20	14.36
22	16.09
24	17.87
26	19.70
28	21.58
30	23.50

1. Batch refining

In the "dry" method, liquid oil or fat is mixed with alkali, heated under agitation, and then the soapstock is allowed to settle by gravity. The upper oil layer is sucked off and the bottom soap layer dropped into another tank. Considerable "art" is involved based on long experience with the oil being handled. The process is slow and losses higher than by centrifugal methods.

In the "wet" method the heated oil is mixed with alkali and the soapstock washed down with sprays of hot water directed at the surface of the oil. This method is practised most where the crude oil is relatively high in free fatty acid and produces soft soapstock. Again, considerable "art" is involved and losses are higher than in a continuous process.

2. Conventional continuous alkali refining

This method differs from batch refining in that mixing of oil and alkali is continuous and the separation of refined oil from soapstock is carried out in centrifuges, thereby reducing the contact time between the oil and alkali (and thus the possibility of saponification of neutral oil) while also effecting a very efficient separation of oil from soapstock. It is at its best when large quantities of the same oil are to be refined. The first process was developed in the 1930s in the United States where two oils, cottonseed and soybean oil, dominated the market and made long runs on a single oil common.

In the early developments, the oil–alkali mixture was separated under atmospheric conditions in small open tubular bowl centrifuges (Fig. 12.1) rotating at high speeds to develop forces of the order of 13,000 g. The heavy phase outlet diameter was varied by means of discharge ring dams located at the top of the bowl. To change ring dams temporary shutdown and dismantling was required. In addition, only small concentrations of solids could be handled, necessitating frequent interruptions for cleaning in refining oil. As time passed, machines were made larger and still later a disc bowl type was developed (Fig. 12.2). This operated under hermetic conditions, rotating at a lower speed but developing centrifugal force up to 7000 g. In these machines the oil–soapstock interface could be adjusted while running merely by changing the pressure on the oil leaving the machine. A further development was a self-cleaning bowl, opening and closing by hydrostatic pressure and making possible discharge of accumulated solids at will (Fig. 12.3). Machines of this type now have capacities of 25,000–50,000 lb per hour as compared to about 1000 lb per hour for the early centrifuges. The tubular bowl machines are still used in many cases where handling of solids is no problem.

In a common continuous process (Fig. 12.4), a large supply of crude oil (perhaps a day's production) in a "day tank" is continuously fed to the system where oil and caustic are accurately proportioned for a smooth non-pulsating flow of product to mixers. The latter are normally three to four horizontal compartmented or vertical disc and doughnut units connected in series as needed for required contact time. The oil–caustic mixture is now heated to get a good "break" (74°C, 165°F). Close temperature control is essential. The hot mixture goes to a primary centrifuge where the oil and soapstock are separated. The refined oil is then heated to 74–82°C (165–180°F), mixed with about 10–20% of soft hot water (88–93°C, 190–200°F) low in iron and copper and the washed oil separated in a high capacity centrifuge. A second wash is sometimes needed, although one wash will ordinarily remove about 90% of the soap. The washed oil is now spray dried by flashing into a continuous vacuum dryer.

Some oils, especially soybean and rapeseed, contain gums (phosphatides) which are difficult to remove by repeated alkali treatment and when present tend to cause poor taste and lower oxidative stability in the finished deodorised oil. Their removal is facilitated by adding 0.03–0.15% of concentrated phosphoric acid to the crude oil before alkali refining. Of course, enough additional

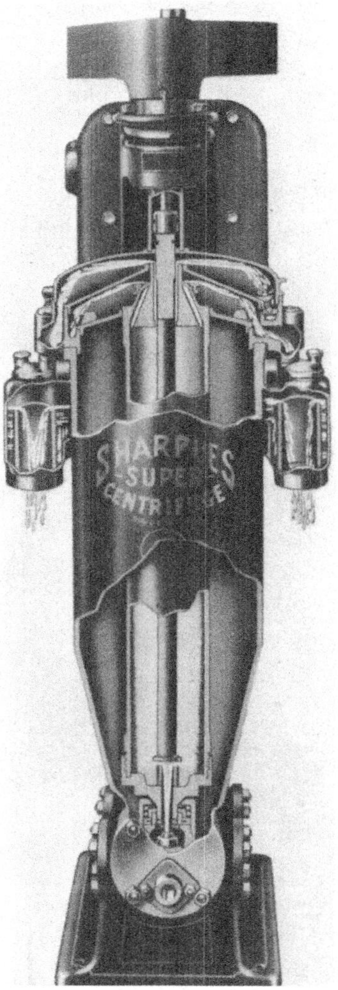

FIG. 12.1. Cut-away view of tubular bowl centrifuge. (Courtesy Penwalt Sharples Corporation, Oak Brook, Illinois.)

alkali must be added to neutralise the phosphoric acid when the oil is refined. Another advantage of this treatment is that the phosphoric acid tends to remove metal ions in the oil through its sequestering ability.

3. Zenith process

One continuous process that does not involve centrifuges is the Zenith, practised in Europe to some extent. In this method the crude oil is first treated with phosphoric acid and then neutralised as it rises as droplets by gravity through a weak lye column. Emulsions are minimised and water washing of the neutralised oil is unnecessary.

4. Miscella refining

When alkali refining is done at an oil mill, the crude oil can be treated while still dissolved in solvent. This so-called "miscella refining" gives minimum refining loss and a good quality refined oil. It is uneconomical to add solvent at a refinery just for the advantages gained by reaction in a solvent.

In miscella refining an oil–solvent mixture of 30–70 % concentration (commonly 40–58 %) is intimately mixed with alkali as in a homogeniser to hydrolyse the phosphatides and pigments in the crude oil. Alternatively, chemical additives may be added to the crude miscella followed by a

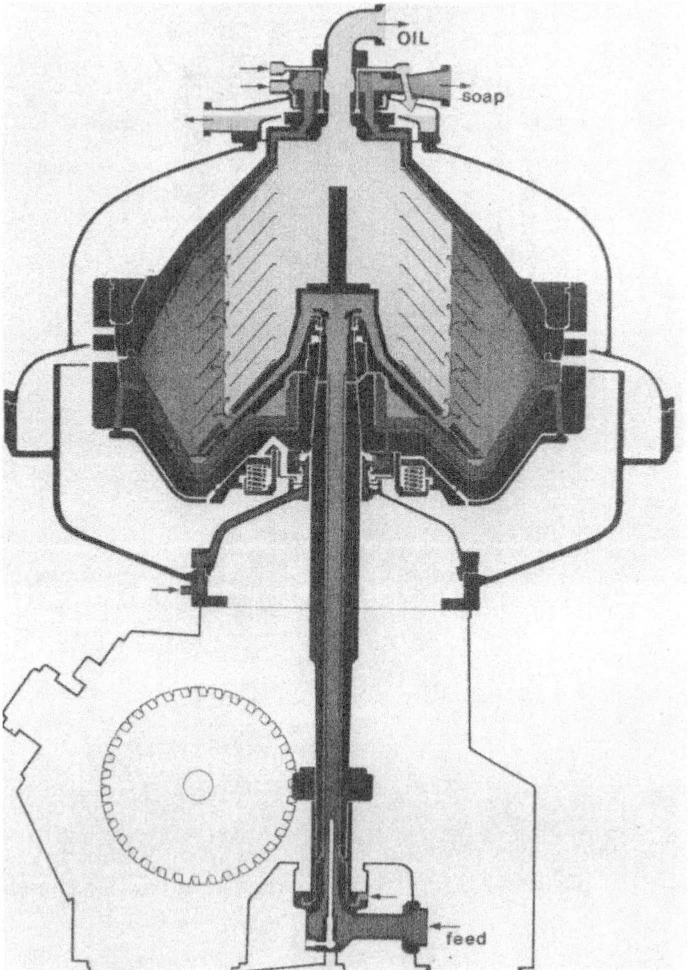

FIG. 12.2. Sectional view of disc bowl centrifuge. (Courtesy of the Sullivan Systems Inc., Tiburon, California.)

FIG. 12.3. DeLaval self-cleaning refining centrifuge (SRPX-317) (Courtesy of the DeLaval Separator Company, Poughkeepsie, New York.)

short mixing period before the addition of the alkali. When a homogeniser is used, caustic soda is added to provide about 0.3 % excess and the homogenised mixture is heated to about 65°C (149°F) to melt the soapstock, then cooled to 45°C (120°F) and separated in hermetic centrifuges. The neutralised oil is then recovered by stripping off the solvent (hexane).

An additional benefit of miscella refining is that, if desired, the oil can be winterised while still in solvent to separate the more saturated fats from the more unsaturated ones. Also, continuous hydrogenation of the refined miscella is possible.

5. Soapstock handling

An environmental problem arising from alkali refining is the disposal of the soapstock which results from the action of the alkali on the free fatty acids. Before the advent of synthetic detergents

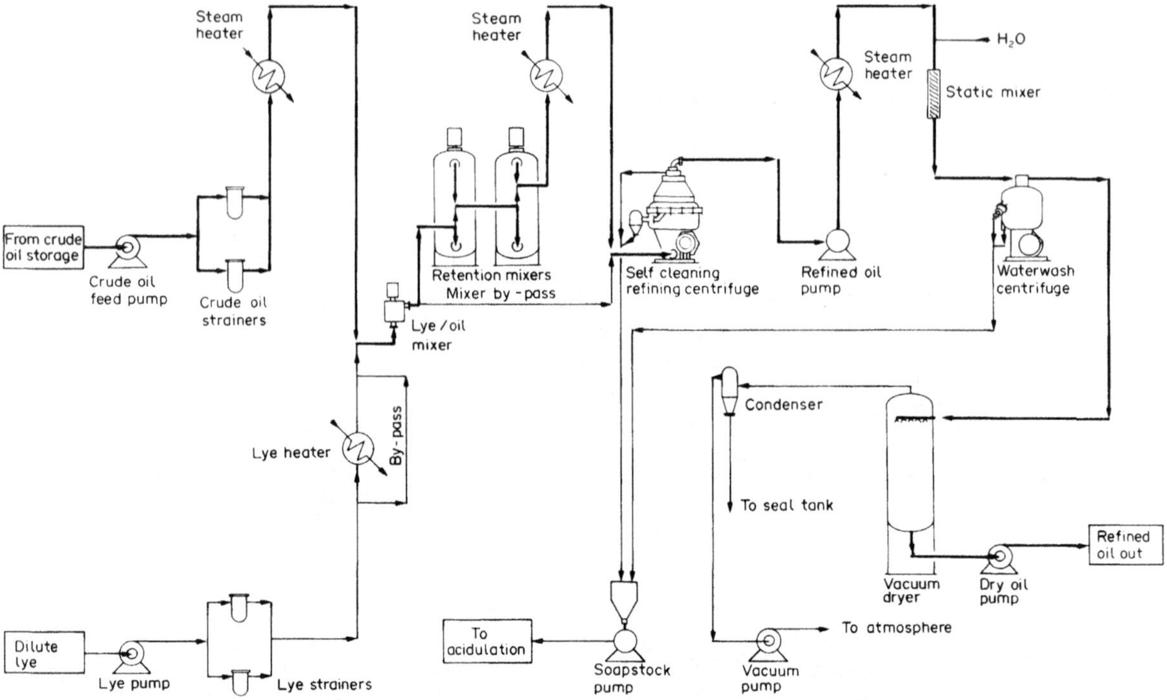

FIG. 12.4. Basic flow diagram of caustic refinery. (Courtesy of the Sullivan Systems Inc., Tiburon, California.)

much of the soapstock was utilised in such products as laundry soaps. Nowadays this use is very limited and soapstock is commonly acidified with an inorganic acid like sulphuric to produce free fatty acids for feeds and industrial use. Corrosion is a problem here as is also the disposal of the acid waste water which normally must be neutralised before leaving the refinery. A recent development involves neutralising and drying the soapstock to produce a neutral product which, it is claimed, possesses the same feed efficiency and gives the same rate of weight gain of chickens equal to that obtained with a commercial feed fat at the same level.[1]*

C. OTHER REFINING METHODS

1. Physical refining

It has been known for a long time that oils containing very little phosphatides could be freed of free fatty acids by a simple vacuum distillation (steam refining). In the case of some high free fatty acid oils this treatment avoided very large refining losses due largely to emulsification. In most cases, distillation down to a free fatty acid level of 0.2–0.5 % was followed by conventional alkali refining. Recently, it has been claimed that alkali refining can be avoided entirely by treating the crude oil with phosphoric acid to remove phosphatides, adding bleaching earth to take up the acid, heating to 163–260°C (325–500°F), cooling, filtering, and finally submitting the oil to

* Superscript numbers refer to References at the end of Chapter.

deodorisation.[2] This process has reportedly worked well on palm oil (low in phosphatides), but at this time remains to be demonstrated on soybean oil (higher in phosphatides) on a plant scale.

2. Degumming

Degumming is only a partial refining, since free fatty acid is not reduced and even the gums are not completely removed, but its widespread use merits some discussion. Crude soybean oil contains about 2% phosphatides ("gums"). When heated to 70–80°C (158–176°F) and mixed with 1–3% water, the phosphatides hydrate and become oil-insoluble. On a batch basis 30–60 min is required for complete hydration. In a continuous system the time may be reduced to 1 min or less when very good mixing is used. A more complete degumming may be obtained by adding a little concentrated phosphoric acid (0.02–0.5%) to the oil before mixing. This removes most of the so-called "non-hydratable phosphatides" that resist hydration with water alone. The procedure is not recommended where lecithin is to be recovered for sale, since it adversely affects the lecithin quality.

In the production of lecithin, minimum water is used, since less drying is then required to obtain a final moisture under 1%. Wet gums are commonly vacuum dried at 115°C (239°F) and then cooled to about 50°C (122°F) before shipment or blending. This final product is plastic, but can be made fluid by adding free fatty acids or esters. The National Soybean Processors Association[3] has issued specifications for plastic and fluid unbleached, single, and double bleached lecithin (Table 12.2).

TABLE 12.2. *Soybean Lecithin Specifications*

Grade:	Fluid natural lecithin	Fluid bleached lecithin	Fluid double-bleached lecithin
Analysis			
Acetone insoluble, min.	62%	62%	62%
Moisture, max[a]	1%	1%	1%
Benzene insoluble, max.	0.3%	0.3%	0.3%
Acid value, max.	32	32	32
Colour, Gardner, max.[b]	10	7	4
Viscosity, poises, at 77°F, max.[c]	150	150	150

Grade:	Plastic natural lecithin	Plastic bleached lecithin	Plastic double-bleached lecithin
Analysis			
Acetone insoluble, min.	65%	65%	65%
Moisture, max.[a]	1%	1%	1%
Benzene insoluble, max.	0.3%	0.3%	0.3%
Acid value, max.	30	30	30
Colour, Gardner, max.[b]	10	7	4
Penetration, max.[d]	22 mm	22 mm	22 mm

[a] By toluene distillation for 2 hr or less (A.O.C.S.Ja. 2–46).
[b] On a 5% solution in mineral oil.
[c] By any appropriate conventional viscosimeter, or by A.O.C.S. Bubble Time Method Tq. 1A-64, assuming density to be unity. Fluid lecithin having a viscosity less than 75 poises may be considered a premium grade.
[d] Using precision cone 73525, penetrometer 73510; sample conditioned 24 hr at 77°F.

Soybean oil can also be degummed by acetic anhydride, ammonia, and various organic and inorganic acids, but water with or without phosphoric acid is most common.

3. Miscellaneous refining methods

Sodium carbonate ("soda ash") was popular at one time, since it neutralised free fatty acids without saponifying any oil. In a second step, stronger alkali could be used for colour reduction, etc. Foaming problems and the equipment required have caused this method to be discarded in most refineries. Likewise, refining by liquid–liquid extraction using furfural, propane, organic alkalis, etc., has not gained acceptance for a variety of reasons.

D. MEASURING REFINING LOSS

The difference between the weight of crude and refined oil is the simplest measurement of refining loss; however, this is seldom practical and more indirect methods are used. These include comparing the weight of neutral oil produced to the calculated neutral oil present in the crude as determined by laboratory methods ("refining efficiency") or by comparing the plant loss to the free fatty acid in the crude. A more recent technique is the continuous metering of crude and refined oil to give an immediate measure of refining loss. With improvements in liquid flow metering this method is rapidly gaining favour, especially since it tends to pinpoint current refining loss and increase operator awareness. Very small increases in refining loss may represent large monetary loss in high volume refineries.

REFERENCES

1. R. E. Beal, V. E. Sohns, and H. Menge, *J. Amer. Oil Chemists' Soc.* 1972, **49**, 447.
2. F. Taylor, U.S. Patent 3,895,042 (1975).
3. National Soybean Processors Association, *Yearbook and Trading Rules*, 1978–1979, Washington, D.C.

GENERAL REFERENCES

Brae, Ben, "Degumming and Refining Practice in Europe", *J. Amer. Oil. Chemists' Soc.* 1976, **53**, 353.
Carr, Roy A., "Degumming and Refining Practice in the United States", *J. Amer. Oil. Chemists' Soc.* 1976, **53**, 347.
Cavanagh, George C., "Miscella Refining", *J. Amer. Oil Chemists' Soc.* 1976, **53**, 361.
Duff, A. J., "Automation in Vegetable Oil Refineries", *J. Amer. Oil Chemists' Soc.*, 1976, **53**, 370.
Sullivan, Frank E., "Steam Refining", *J. Amer. Oil Chemists' Soc.* 1976, **53**, 358.

CHAPTER 13

Bleaching

Refined oil may still contain small amounts of impurities that must be removed to produce an oil of colour and flavour acceptable to the finished oil buyer. Also, a hydrogenated oil usually requires a light bleaching treatment to remove traces of hydrogenation catalyst. Colour removal is the most obvious result of bleaching, but other effects may be just as important. Many processors give their oil a good bleach with activated earth, even though a mild bleach with neutral earth plus the "heat-bleach" encountered at deodorisation temperatures will bring the colour to the level necessary to meet their finished product specifications. This makes for premium quality oil.

Among the impurities more or less completely removed by bleaching are residual soaps (from alkali refining), chelated pro-oxidant metals, sulphur compounds, peroxides, and traces of aldehydes and ketones arising from decomposition of peroxides. The latter may be in concentrations of a few hundred parts per million in oil going to deodorisation. In soybean oils which were refined and deodorised under commercial conditions in the same plant, bleached oil was shown to produce a more bland and more flavour-stable product than unbleached oil from the same refined oil[1]* (Table 13.1). Also, peroxides developed more slowly.

TABLE 13.1. *Quality of Bleached vs. Unbleached Soybean Oils*[a,b]

	Flavour score (peroxide value)	
Factor	Bleached	Unbleached
Initial flavour[c]	7.9 (0.8)	6.7 (0.5)
Aged flavour	6.4 (1.8)	5.2 (4.0)
8-hr AOM peroxide value	2.2	20.0

[a]From reference 1.
[b]Both oils were refined and deodorised in a commercial plant, but bleaching was omitted for one sample. Oxidation was minimised and a metal scavenger was added to both deodorised oils.
[c]Based on a 1–10 flavour scale, with 10 being the highest rating. Peroxide values, in parenthesis, are additional evidence since a low peroxide value generally means greater stability.

A. COLOUR STANDARDS

Most non-crude edible oils have their colours determined by the Wesson method, which involves matching the colour of a $5\frac{1}{4}$ in column of melted oil or fat against standardised yellow and red

* Superscript numbers refer to References at end of Chapter.

Lovibond glasses. Since the yellow colour affects the red reading, Trading Rules and industry practice usually specify the ratio of yellow to red (Y/R) for different kinds and types of oils. For example, 35 yellow (expressed 35 Y) is specified for refined cottonseed oil and refined peanut oil but 70 Y for refined soybean oil. A 10:1 yellow to red ratio is common for deodorised oils up to 3.5 red (3.5 R). Although certain oils are sold darker by trade preference, most oil used in edible products has a Lovibond red colour of about 2.5 or less. Many shortenings are as light as 1.0 R or less. Bleaching generally may be said to reduce refined oil colour from the 4–9 red range to the 1.5–2.5 range. The visual determination of oil colour is quite subjective, and in view of the economic importance of colour readings, a photometric method has been developed[2] and is gradually meeting increased acceptance. It is based on measuring the optical density of the oil at 460, 550, 620, and 670 mμ in a 21.8 mm cell:

$$\text{Photometric colour} = 1.29A_{460} + 69.7A_{550} + 41.2A_{620} - 56.4A_{670}$$

where A is the optical density or absorbance.

There are also several newer visual colour methods instrumented for ease of operation and accuracy of results, as well as the Gardner colour standards for drying oils and the FAC Colour Standard for Tallows and Greases.[3]

Crude soybean oil may have a greenish cast, which is undesirable for finished oil. Accordingly, the crude oil is compared to two standard solutions (A and B) and refined and bleached oil discounts are heavier for oils whose crude colours are equal or darker than standard A but lighter than standard B.[4]

B. METHODS OF BLEACHING

1. Heat

Some pigments, for example carotenes, become colourless if heated sufficiently; however, this leaves the pigment molecules in the oil and may have an adverse effect on quality. Also, if contact with air occurs, coloured degradation products, such as chroman-5,6-quinones from γ-tocopherol present, may be formed. These are very difficult to remove.

2. Chemical oxidation

Some pigments such as the carotenoids are made colourless or less coloured by oxidation; however, such oxidation invariably affects the glycerides and destroys natural antioxidants. Consequently, it is never used for edible oils but is restricted to oils for technical purposes, such as soapmaking.

3. Adsorption

Bleaching earths possess a large surface that has a more or less specific affinity for pigment-type molecules, thus largely removing them from oil without damaging the oil itself. Adsorption is the method commonly used for bleaching edible oils.

C. BLEACHING BY ADSORPTION

1. Theory

A fundamental empirical equation was developed by Freundlich[5] correlating the capacity of the adsorbent with the residual concentration of the material adsorbed. It is:

$$\frac{x}{m} = Kc^n$$

where x = amount of solute (pigment) adsorbed,
m = amount of adsorbent (bleaching earth),
c = amount of residual solute,
and K and n are constants.

K is a measure of the decolourising power of the adsorbent for the pigment involved and n indicates the range of decolourisation where the adsorbent is most effective.

Using the Freundlich equation, bleaching earths can be compared to one another under various bleaching conditions.

2. Bleaching earths

The bulk of natural bleaching earths are complex silicates from which some of the aluminium ions have been leached by prolonged weathering. Variable amounts of magnesium, calcium, iron, and sodium are present. Activation is achieved by treating certain inactive earths with mineral acids, usually hydrochloric or sulphuric. This greatly increases the void spaces in the particles as well as the total surface area, causing increased bleaching activity. It also increases oil retention. In general, activated earths are used for the more difficult bleaching jobs. Activated carbon is also a very effective bleaching agent but it is expensive and may retain up to 150 % of its weight of oil as compared to 20–30 % for neutral earths and 30–50 % for activated earths. It is usually used in combination with earths as 10–40 % of the total amount.

3. Bleaching conditions

Minimum earth dosage is about 0.2–0.4 % of the weight of oil,[6] but may reach 1.5–2.0 %, depending upon the oil and the activity of the earth. Temperatures are often around 104°C (220°F) for atmospheric bleaching, but somewhat lower (82°C, 180°F) if vacuum is used. There are some differences of opinion with regard to optimum temperatures. With activated earths, temperatures above 95°C (203°F) favour splitting of soap to fatty acids and above about 160°C (320°F) isomerisation of the fatty acids is more likely.[6] A 10–15 min contact time at the top bleaching temperature is common.

Most processors believe that bleaching earths should contain some moisture which is preferably lost while in contact with the hot oil. Suppliers generally offer earths with an optimum moisture content for maximum effectiveness. Oil is customarily dried to a maximum moisture content of less than 0.2 % before bleaching.

Agitation is not critical. It should be enough to provide good contact between the earth and the oil.

Vacuum is an important factor, and atmospheric bleaching is confined mainly to older installations. Bleaching proceeds well under atmospheric conditions, but with air present some pigments tend to become "fixed" by oxidation. In addition, it has been shown that oxidative and flavour stability are improved by vacuum or inert atmosphere bleaching. In general, contact of oil with air during and after bleaching and before deodorisation should be avoided as much as possible.

Finely ground earths have more surface area and hence more colour-absorbing power. On the other hand, as particle size becomes very small, filtration rate is seriously reduced and oil retention increases. One commercial activated earth has 100, 94, and 75% passing through the 100-, 200-, and 325-mesh Tyler screens.[7]

D. BATCH BLEACHING

In atmospheric bleaching, open cylindrical cone-bottom kettles are used, holding up to 29,300 kg (60,600 lb) of oil. These are equipped with steam coils and agitators. As the oil is heated to about 71–82°C (160–180°F) the bleaching earth is added, perhaps as a slurry, and the heating continued to 100–104°C (212–220°F). After 15–20 min at the top temperature, filtration is started and the oil is recycled until the filtrate is clear and brilliant. It is then cooled and sent to storage or the next operation. Many different types of filters are available for ease of operation and reduced labour. Press cake is usually recovered by blowing with air or nitrogen, followed by dry steam. In a few large installations, the press cake is solvent extracted to recover the oil. Costs and safety problems mitigate against this procedure in most plants.

So-called "press bleach" is the name given to the added bleaching action that occurs in the filter press where the concentration of adsorbent is very high in relation to oil and may be part of the reason why plant bleaching is invariably superior to laboratory results.

In vacuum bleaching, closed cylindrical vessels with dished ends are employed. A vacuum of 26–28 in of mercury is common. The incoming oil is sprayed into the vessel for immediate deaeration. The adsorbent is added before the oil is heated or at or near the bleaching temperature. After about 15–20 min agitation at bleaching temperature, the charge is cooled to 71–82°C (160–180°F), filtered, and finally cooled to 38–66°C (100–150°F) before exposure to air.

E. CONTINUOUS BLEACHING

As shown in the flow diagram (Fig. 13.1), the crude oil is fed continuously at a regulated rate into the slurry tank (3). Activated clay and filter aid are continuously fed from the clay bin (4) into the slurry tank (3) at a controlled rate adjusted to suit the characteristics of the oil. The slurry of oil, clay, and filter aid is pumped continuously to the vacuum bleacher (5) and sprayed into the head space to obtain complete deaeration prior to heating.

The deaerated slurry is heated to the desired bleaching temperature under automatic temperature control by internal steam coils and the bleacher is maintained under vacuum by a steam jet ejector system (7). The vacuum bleacher is compartmented to provide two stages of mixing for intimate contact and adequate time. The slurry is pumped from the vacuum bleacher

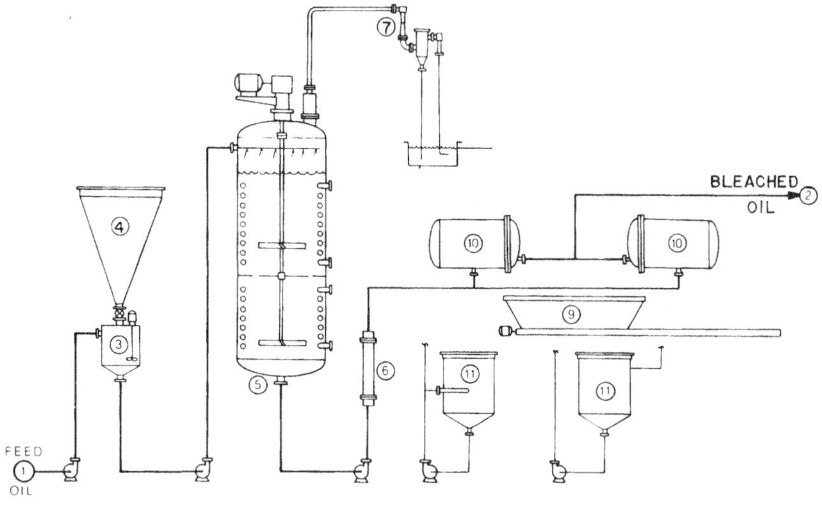

FIG. 13.1. Continuous vacuum bleaching system. (Courtesy of EMI Corporation, DesPlaines, Illinois.)

through a cooler (6) and into one of the two filters (10), which are provided for alternate use. The filtered pretreated oil is discharged continuously in the steam refining deodorising process or to intermediate storage.

When a filter reaches its cake capacity, the slurry flow is diverted to the other filter. The oil is discharged from the out-of-service filter; and the cake is blown with steam, emptied into the cake chute (9), and conveyed to the discharge point. The filter then is precoated, using the precoat tank (11) and pump, in preparation for the next filter cycle. Steamings from the filter are collected in the steamings tank (8) and reworked into the slurry tank (3).

F. BLEACHING IN SOLVENT

Like refining, bleaching can also be carried out with the oil dissolved in a solvent, with excellent results. This, however, is done only where a solvent extraction plant is on the premises to supply the oil before the solvent has been removed. It is not economical to dissolve oil in solvent just to refine and/or bleach.

REFERENCES

1. J. C. COWAN, *J. Amer. Oil Chemists' Soc.* 1966, **43,** 300A, 302A, 318A.
2. *American Oil Chemists Society Official and Tentative Methods*, Champaign, Illinois, 1974.
3. V. C. MEHLENBACHER, *The Analysis of Fats and Oils*, The Garrard Press, Champaign, Illinois, 1960.
4. *National Soybean Processors Association Yearbook and Trading Rules*, Washington, D.C., 1978–1979.
5. H. FREUNDLICH, *Colloid and Capillary Chemistry*, translated from the German edition by H. S. Hatfield, Methuen and Co., London, 1926.
6. H. B. W. PATTERSON, *J. Amer. Oil Chemists' Soc.* 1976, **53,** 339.
7. L. L. RICHARDSON, *J. Amer. Oil Chemists' Soc.* 1978, **55,** 77.

GENERAL REFERENCES

GOEBEL, E. H., "Bleaching Practices in the United States", *J. Amer. Oil Chemists' Soc.* 1976, **53,** 342.

NORRIS, F. A., *Bailey's Industrial Oil and Fat Products*, 4th edition, (D. Swern, ed.), John Wiley & Sons, New York, N.Y., 1982, Vol. 2, page 253.

CHAPTER 14

Hydrogenation*

Hydrogenation is the process of adding hydrogen to the double bonds in a fat or oil in the presence of a catalyst, usually nickel. It was discovered by Normann and patented in 1903. This discovery represented a major advance in the processing of oils. Hydrogenation accomplishes two things: it increases both the melting point of the oil or fat and the resistance to oxidation and flavour deterioration. Before the advent of hydrogenation relatively expensive meat fats were used as shortening, alone or blended with smaller amounts of liquid oils. The amount of liquid oil (cottonseed, soybean, etc.) was limited by the need to make a plastic edible fat. In the United States especially, a large production of cottonseed oil became suitable for use as shortening through the use of hydrogenation. Later soybean oil followed the same path. Actually, the whole growth of the soybean industry in the United States paralleled the growth of hydrogenation of food oils. Now liquid oils can be used as they are or in any degree of hardening through hydrogenation.

Dual purpose cooking/salad oils may be made by a light hydrogenation so as to remain liquid but have improved flavour stability. Frequently, a winterisation follows the hydrogenation so that unsaturation can be reduced more without losing the liquid character of the finished oil, especially at refrigerator temperatures. Shortenings, margarines, confectionary fats, and other specialised food fats usually involve a greater degree of hydrogenation and/or a mixture of hydrogenated products.

A. THE REACTION

Hydrogenation of oils and fats involves the addition of hydrogen (present as a gas) to unsaturated bonds in the fatty acid chain, causing them to become saturated by the addition of 1 mole of hydrogen to each double bond, represented by:

$$-CH=CH- \longrightarrow -CH_2-CH_2-$$

Actually, the reaction is much more complex than a simple addition of hydrogen and is still under intensive study (see Chapter 6). A catalyst is always required and this is usually nickel that has been chemically reduced with hydrogen to an active state and dispersed in a protective medium of hardened edible oil or tallow at a 25% level for ease of handling.

Although processors are interested primarily in the saturation of double bonds in fatty acid chains, the hydrogen gas may also react with non-glyceride materials present such as carotene. This

* See also Chapter 6.

is normally not a disadvantage. A serious problem is the selectivity of the reaction. One would usually prefer that hydrogen first completely saturate only one double bond in a polyunsaturated system before attacking the diunsaturated and finally the monounsaturated components. In this way linolenic acid present would be changed completely to linoleic before any linoleic was hydrogenated to oleic, and correspondingly all the linoleic would be hydrogenated to oleic before any oleic was converted to the saturated stearic acid. If this were the case, the linolenic acid in soybean oil could be hydrogenated to linoleic acid and no further, thus producing a linolenic-free liquid oil of increased oxidative and flavour stability. Actually, all double bonds are attacked to some extent, the relative hydrogenation rates being about 12.5, 7.5, and 1 for non-selective conditions. That is, linolenic acid is hydrogenated 12.5 times and linoleic 7.5 times as fast as oleic. Under so-called "selective" conditions, with a nickel catalyst, the ratios are still only about 100, 50, and 1. This means that even under the best commercial conditions, using a nickel catalyst, linolenic acid cannot be removed without some loss of linoleic acid. With special catalysts as, for example, copper, it is possible to hydrogenate soybean oil to a product that is still liquid at room temperature and contains less than 1% linolenic acid. Copper catalysts, however, hydrogenate much slower than nickel, are more sensitive to poisoning, and must be completely removed from the treated oil to avoid the very deleterious effect of copper on flavour stability. To date, these disadvantages have prevented large-scale use.

In addition to relative reaction rates for double bond saturation, other changes may occur during hydrogenation. One is conversion of *cis* double bonds to the higher melting *trans* forms. This stereo-isomerisation always accompanies increased selectivity. It is undesirable in that it increases melting point without decreasing unsaturation and introduces an isomeric structure not originally present. On the other hand, it may produce very desirable physical properties, as for example in coating fats. Still another change accompanying hydrogenation may be the shifting of double bond locations from their original to other positions along the chain. This is the result of the reaction mechanism. Conjugation of double bonds also occurs to some extent.

B. SELECTIVITY

Hydrogenation rates can be expressed as follows:

$$\text{Linolenic} \xrightarrow{K_{Ln}} \text{Linoleic} \xrightarrow{K_{Lo}} \text{Oleic} \xrightarrow{K_O} \text{Stearic}$$

Knowing the composition of starting soybean oil and hydrogenated oil (by GLC) plus the reaction time, the pseudo first order reaction rate constants K_{Ln}, K_{Lo}, and K_O can be calculated by computer or by using published graphs.[1]* For example, if $K_{Lo} = 0.159$ and $K_O = 0.013$, then $K_{Lo}/K_O = 0.159 \div 0.013$, or 12.2. This means that the linoleic acid hydrogenates 12.2 times as fast as the oleic. It is called the selectivity ratio (SR). Values of 2–70 have been reported for this in the literature. Selectivity is influenced by operating conditions, catalyst, and whether or not the catalyst has been used previously.

* Superscript numbers refer to References at end of Chapter.

C. HYDROGENATION REQUIREMENTS

1. Hydrogen gas

The hydrogen used should be dry and as pure as possible. It can be produced by the steam-hydrocarbon, hydrocarbon reforming, or other methods. Pure liquid hydrogen can also be obtained commercially. Due to its explosive nature, special safety precautions are essential.

2. Oil

The oil to be hydrogenated normally must be refined, bleached, very low in soap (under 25 ppm), and dry. This is to prevent poisoning of the catalyst and reduced selectivity. Free fatty acids, soap, and water can all act as catalyst poisons that reduce catalyst activity and selectivity.

3. Catalyst

Most refineries purchase their catalyst (nickel) from suppliers who offer several types for specific uses, with physical characteristics designed to promote good activity and ease of filtration. Catalyst activity can be checked in the laboratory by an American Oil Chemists' Society procedure,[2] which compares the test catalyst to a standard one in hydrogenating soybean oil to 80 IV under specific conditions.

As mentioned, catalysts are susceptible to poisoning by such things as sulphur compounds, fatty acids, soaps, and phosphatides in the oil. Recent work indicates that poisoning may be minimised, in at least some cases, by the addition of activated bleaching earth.[3] Good catalysts may be used more than once, although selectivity will decrease and slightly more may be required to compensate for reduced activity. New catalysts are commonly used to prepare critical base stocks, with used catalysts reserved for fully hydrogenated products where selectivity is unimportant.

D. PROCEDURE

Oil is pumped into a pressure vessel called an autoclave, convertor, or hydrogenator. This must be equipped for good agitation, means for heating and cooling, and with suitable openings for oil, hydrogen, and sampling. Vacuum is applied on the headspace and heating started. Catalyst is weighed into a catalyst mix tank, slurried with a small amount of oil, and when the reaction temperature is reached the catalyst is pumped into the reactor, mixed, and then hydrogen is added to the desired pressure. In the earlier installations where the hydrogen contained quantities of inert gas which could be vented, the hydrogen was recirculated through the oil. The "dead-end" convertor, usually preferred now, allows the hydrogen to accumulate in the headspace above the oil with recirculation provided by the vortex action of the agitator. During hydrogenation, water is circulated through cooling coils to counteract the heat of reaction since hydrogenation is an exothermic reaction, and later for cooling prior to filtration. The general course of the reaction may be followed by the decrease in refractive index (RI); however, congeal points and/or Solid Fat

Index may be needed for adequate information. Also, with iodine values (IV) below 90–95, RIs may be misleading since *trans* isomers formed will make the oil harder than the RI indicates.

After cooling, the hydrogenated oil is pumped to a precoated "black press" and recirculated until clear and brilliant. To remove any nickel passing the filter, citric acid is commonly mixed with the charge in the bleach tank and the batch then pumped through a post-bleach filter and a polishing filter utilising paper as the filter medium.

E. EFFECT OF PROCESS CONDITIONS

An understanding of the effects of varying process conditions is helped by attempting to visualise what must occur during hydrogenation. Hydrogen, dispersed as bubbles, has to dissolve in the oil to reach the catalyst surface. The bubbles are surrounded by a stagnant oil layer as they move through the oil in turbulent movement, and the dissolved hydrogen must pass through this by molecular diffusion. When the gas reaches the stagnant oil surrounding the catalyst particles, it diffuses to the nickel surface on which it is adsorbed as hydrogen atoms. It is now ready to react with unsaturated molecules.

Initially, poly- and monounsaturated acids will be bonded to the nickel surface by one double bond, i.e. by two carbon atoms. If hydrogen coverage is high, fast hydrogenation will occur on the adsorbed double bond and selectivity will be low; however, if hydrogen coverage is low, an adsorbed linoleic acid moiety can have time to split off a hydrogen atom from its reactive $-CH_2-$ group and thus become bonded to the nickel with three carbon atoms. The second double bond is now close to the nickel surface and can be bonded to it. The adsorbed linoleic acid may now take up a hydrogen atom yielding an adsorbed form of 9,11- or 10,12-octadecadienoic acid. This hydrogenates rapidly with some *trans* acid formation.

Increased hydrogen pressure, therefore, can be expected to reduce selectivity and *trans* acid content but increase reaction rate, while reduced hydrogen pressure will tend to increase selectivity and *trans* acid content and to decrease reaction rate. Increased temperature decreases dissolved hydrogen near the catalyst surface, acting like decreased hydrogen pressure and with the same effect.

Increased agitation creates a finer dispersion of hydrogen bubbles in the oil making the concentration of dissolved gas larger and having the same effect as increased hydrogen pressure.

Increased catalyst concentration provides more room for fatty acid adsorption, permitting more selectivity (and *trans* acids) while also increasing reaction rate.

Table 14.1 summarises the effect of processing conditions on selectivity, *trans* acid formation, and hydrogenation rate, using a nickel catalyst.

Typical conditions for selective and non-selective hydrogenation are shown in Table 14.2.

Under non-selective conditions more saturated fat is produced at the same hydrogen absorption or IV decrease. This is shown in Fig. 14.1, which compares the Solid Fat Index (SFI) vs. temperature for selective and non-selective conditions.

To appreciate the importance of the Solid Fat Index one must understand that plastic fats, such as shortenings, are not completely solid but consist of a mass of very small crystals of solid fat enmeshed in liquid oil. The ratio of liquid oil to solid fat varies with several factors, including temperature. As a fat is heated, increasing amounts of liquid oil are produced as the solid glycerides melt. If this occurs over a wide temperature range the fat is said to have a wide plastic range. A margarine oil, for example, should be firm at refrigerator temperature (7–10°C, 45–50°F) but liquefy quickly at mouth temperature. Similarly, the workability and creaming ability of a

TABLE 14.1. *Effect of Processing Conditions on Hydrogenation*

	Selectivity	*Trans* acids	Reaction rate
Increased temperature	+	+	+
Increased pressure	−	−	+
Increased agitation	−	−	+
Increased catalyst concentration	+	+	+

TABLE 14.2. *Suggested Hydrogenation Conditions with Nickel Catalyst*

	Pressure	Catalyst	Temperature
Selective hydrogenation	5–14 psig	0.05 %	177°C (350°F)
Non-selective hydrogenation	50 psig	0.05 %	121°C (250°F)

shortening at any given temperature is largely related to the content and type of solid glycerides at that temperature. Solid Fat Index is the most widely used method for determining the consistency of fats.

The analysis for SFI is based on the fact that as a fat melts there is an increase in volume due to phase transformation and this increase is many times the volume increase due to thermal expansion (Table 14.3). Thus, when volume changes are plotted against temperature a melting dilation curve is obtained which shows the solids present at each temperature. This is an arbitrary measurement, but can readily be correlated with mouth feel and plasticity. SFIs are commonly reported at 10°, 21.1°, 26.6°, 33.3°, and 37.8°C (50, 70, 80, 92, and 100°F). A margarine may have an SFI of 15–28 at 10°C (50°F) and 1.5–4 at 33.3°C (92°F). Too low an SFI at 10°C is detrimental to plasticity and too high an SFI at 33.3°C (92°F) leaves a waxy sensation in the mouth. All solids should melt quickly at mouth temperature.

SFIs for a few margarines and shortenings are shown in Table 14.4.

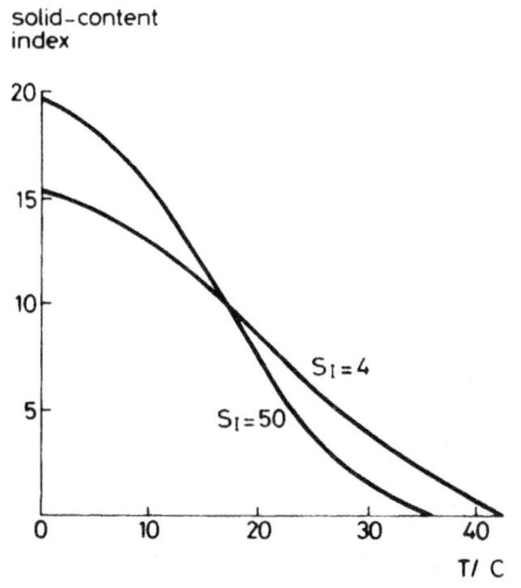

FIG. 14.1. Solids content versus temperature for soybean oil hydrogenated under conditions of high selectivity ($S_I = 50$) and low selectivity ($S_I = 4$), from J. W. E. Conen, *J. Amer. Oil Chemists' Soc.* 1976, **53**, 386.

TABLE 14.3. *Melting Dilations and Expansivities of Some Synthetic Triglycerides*

	Melting dilation ml/g	Expansivity ml/g/°C	
		Solid	Liquid
Tripalmitin	0.1619	0.00022	0.00092
Tristearin	0.1671	0.00023	0.00092
Trielaidin	0.1180	0.00018	0.00087
Triolein	0.083	0.00030	0.00082
Oleodistearin	0.1106	0.00030	0.00092

TABLE 14.4. *Some Typical SFI Values*

	10°C 50°F	21.1°C 70°F	26.6°C 80°F	33.3°C 92°F	37.8°C 100°F
All hydrogenated shortening					
Standard	27.1	18.4	16.9	11.8	8.7
High stability	43.8	27.5	22.0	11.3	4.7
"Stick" margarine	27	16	12	3–5	0
Harder component	66	59	57	43	27
Softer component	18	7	2	0	0
Soft tub margarine	13	8	6	2	0
Bakers margarine	27	18	16	12	8

F. CONTINUOUS HYDROGENATION

In batch hydrogenation the pressure vessel is not utilised efficiently because it is used both to heat the oil and as a holding tank prior to filtration. In a continuous system there is an obvious heat economy, since part of the heat needed to preheat the entering oil can be provided by a heat interchange with the exit oil. Also, once optimum operating conditions are established it would appear easy to continue to make a very uniform product. Obviously, the advantages are reduced or even eliminated unless long runs are possible without change of feedstock or desired end product. This has greatly hampered the use of continuous hydrogenation, although it is used commercially to some extent. At least two such plants are operating in the United States.

In practice, most continuous hydrogenation installations are cascades of a number of stirred tank reactors. If only one tank is used, there is a wide spread in residence times. To avoid this a number of reactors (5–6) are connected in series, giving a narrowed time distribution curve. Even so, some of the charge stays in the system longer and some shorter than the average residence time. Thus, the end product will contain both over- and under-hydrogenated product, i.e. too much saturated and too much unsaturated. Also, a residence time distribution lowers selectivity and the SFI–temperature curve will be somewhat less steep for a continuous vs. a batch hydrogenated product. Continuous hydrogenation using a packed bed column containing catalyst pellets through which hydrogen and oil are passed, either concurrently or countercurrently, appears to suffer from poor selectivity and inefficient use of catalyst.

One plant using a flow-through pipe with measured amounts of hydrogen added in increments to the oil as it moves through the system has been in operation for several years. It was patented in

1974[4] with rights sold to Dravo. Lurgi also has continuous hydrogenation plants apparently operating on fatty acids but not glycerides.

REFERENCES

1. L. F. ALBRIGHT, *J. Amer. Oil Chemists' Soc.* 1965, **42,** 250.
2. Method Number Ca-17-76 in *Official Methods of the American Oil Chemists' Society*, 1975, Champaign, Illinois.
3. B. DROZDOWSKI and I. GORAJ-MOSZORA, *J. Amer. Oil Chemists' Soc.* 1980, **57,** 171.
4. W. A. COOMBES *et al.*, U.S. Patent 3,792,067 (1974).

GENERAL REFERENCES

ALLEN, R. R., "Hydrogenation", *J. Amer. Oil Chemists' Soc.* 1981, **58,** 166.

HASTERT, R. C. "Practical Aspects of Hydrogenation and Soybean Salad Oil Manufacture", *J. Amer. Oil Chemists' Soc.* 1981, **58,** 169.

CHAPTER 15
Deodorisation

A. INTRODUCTION

The demand for tasteless and odourless fats was largely a by-product of the increasing acceptance of margarine as a substitute for butter. As the demand for neutral animal fats began to exceed the available supply, research was undertaken to remove the relatively strong flavours of vegetable oils and thereby make these oils suitable for incorporation in margarine. In the United States, David Wesson improved on European practice and finally arrived at what is now common deodorising procedure by a combination of high vacuum and high temperatures, essentially vacuum steam distillation in which the undesirable materials are removed from the relatively non-volatile glycerides without damaging the latter. The process also works well on hydrogenated oils and fats used in margarines and shortenings.

A refined and bleached oil is essentially free of phosphatides, is low in free fatty acids ($\simeq 0.1\%$), and has a greatly reduced content of colour bodies. It still contains aldehydes, ketones, alcohols, hydrocarbons, and miscellaneous other compounds derived from the decomposition of peroxides and pigments. The concentration of these minor components is usually no more than 1000 ppm and in a well-processed oil is frequently at the 200 ppm level; however, many of these compounds may be detected in oil by flavour or odour at the 1–30 ppm level and one compound, decadienal, has a reported flavour threshold in water of 0.5 ppb.[1]* The importance of a thorough deodorisation is obvious.

Because of the heat treatment involved in deodorisation, heat-labile pigments, such as the carotenoids, are bleached and the oil becomes lighter in colour. This so-called "heat bleach" is normally a plus, although there is some question as to the effect on oil quality of leaving excessive amounts of these materials in the finished oil as compared to removing them by conventional bleaching.

Deodorisation is based on the finding that the volatile undesirable materials present in refined and bleached oil have partial pressures which presumably equal or exceed those of the common fatty acids, palmitic and oleic. Thus, flavour and odour removal tends to parallel fatty acid reduction and when the free fatty acid has been lowered to 0.01–0.03% the oil is usually bland. In general, the free fatty acid cannot be reduced below about 0.005% because at this point hydrolysis of the oil by the stripping steam is continually producing more free fatty acid.

Deodorised oil is evaluated by flavour, free fatty acid, colour, peroxide value (PV), etc., depending somewhat upon the individual oil. Freshly deodorised soybean oil in the United States must meet the following specifications of the National Soybean Processors Association:[2]

* Superscript numbers refer to References at end of Chapter.

Flavour	— Bland
Colour	— 20 Y/2.0 R maximum
Free fatty acid	— 0.05 % maximum
Clear and brilliant	—
Cold test	— 5½ hr miminum at 0°C
Moisture and volatile	— 0.10 % maximum
Unsaponifiable matter	— 1.5 % maximum
Peroxide value	— 2.0 maximum
Stability	— 8 hr minimum to reach a peroxide value of 35 mg/Kg

Analyses must be performed by specified methods.

"Physical refining" is a process similar to deodorisation. In physical refining, the free fatty acids are distilled off rather than removed by treatment with alkali ("refining"). Certain high acid oils have been processed in this way for years to minimise refining loss. Normally, the free fatty acid is reduced to 0.5–0.8 % and the oil then further processed by conventional caustic refining, bleaching, and deodorisation. Recently, it has been claimed that fully deodorised oils can be obtained without refining by merely degumming completely and then deodorising.[3] To date this process appears to be more satisfactory on oils, like palm, naturally low in gums (phosphatides) because their complete removal is difficult.

B. DEODORISATION THEORY

Deodorisation is essentially a steam distillation under vacuum. At reduced pressures and elevated temperatures the flavour and odour compounds are more volatile. This is helped further by the use of an inert gas, usually steam, introduced in such a manner as to intimately contact the oil.

Based on Raoult's and Dalton's Laws, a mathematical equation can be developed expressing these factors as follows:

$$S = \frac{PO}{EP_vA}\left(\ln\frac{V_1}{V_2}\right), \tag{1}$$

where S = moles of stripping steam,
O = moles of oil,
P_v = vapour pressure of the FFA,
P = total system pressure,
V_1 = initial number of moles of FFA in the oil,
V_2 = final number of moles of FFA in the oil,
E = vaporisation efficiency of the steam,
and A = activity coefficient (to account for the observed deviation from ideal solutions).

Thus, the amount of stripping steam needed to deodorise a batch of oil varies directly with:
(a) size of the oil batch;
(b) absolute pressure in the deodoriser;
(c) logarithmic ratio of the initial to final concentrations of the FFA (and other volatile constituents) in the oil;

and inversely with
(a) vapour pressure of the volatile compounds at the deodorisation temperature;
(b) vaporisation efficiency;
(c) activity coefficient.

For continuous deodorisation the equation may be written:[4]

$$\frac{V_1}{V_2} = 1 + \left(\frac{K}{P}\right)\left(\frac{S}{O}\right) P_v \tag{2}$$

where V_1, V_2 = moles/hr of FFA in the oil entering and leaving the deodoriser,
S = moles/hr of stripping steam,
O = moles/hr of oil,
and K, P, P_v are the same as for equation (1).

This states that for a given flow rate and initial FFA concentration (V_1) the FFA concentration in the deodorised oil (V_2) will decrease as the deodoriser temperature (which controls P_v) and the steam flow rate (S) increase and the deodoriser pressure (P) decreases.

C. VARIABLES IN DEODORISER OPERATIONS

1. Vacuum

Equation (1) shows that the quantity of stripping steam needed is directly proportional to the absolute pressure in the deodoriser. Thus, if the pressure is increased from 5 to 10 mm, the stripping steam needed is doubled. At 20 mm the requirement would be quadrupled. Most deodorisers are operated in the 3–8 mm range, 6 mm being a good standard figure, although some installations operate as low as 1 mm. Three- or four-stage steam ejectors can produce these vacuums routinely. The stripping steam savings at 1 mm pressure normally are not justified by the increased operating cost unless there is a demonstrable quality difference in the deodorised oil. Good vacuum also has an effect in reducing deodorisation time, since the overall stripping steam velocity is not reduced as fast with pressure as is the total amount required. For example, going from 24 to 6 mm pressure reduces required stripping steam to $\frac{1}{4}$ of the previous amount, but the flow rate is only reduced to $\frac{1}{2}$, making it possible to put through the required amount of stripping steam in less time.

2. Temperature

The vapour pressures of the volatile constituents in oil increase rapidly with increases in temperature, so operation at the highest practical temperatures will minimise deodorisation time. Excessive temperatures produce thermal effects that harm oil quality and must be avoided. The top temperature limit appears to be 15–30 min at 274°C (525°F).[5] In general, American processors prefer high temperature and short time while many European plants opt for lower temperatures and longer deodorisation time. The process is time–temperature related. Semi-continuous deodorisers handling refined and bleached soybean oil typically operate at a top temperature of 246–260°C (475–500°F) for 15–40 min at an absolute pressure of 3–6 mm Hg. Heating is normally accomplished with high-pressure steam initially and then Dowtherm A or Therminol VP-1. Both are trade names for a eutectic mixture of diphenyl and diphenyl oxide which boils at about 255°C (496°F) at atmospheric pressure. Only about 16 psig is needed to deodorise at 257°C (500°F). Using steam alone, the required pressure would be about 900 psig. In some installations all the heating is done by the thermal fluids rather than by steam alone or a combination of steam and thermal fluid.

3. Stripping steam

In most deodorisations stripping steam is used for agitation as well as stripping, and allowances must be made for this. Most of the steam should go into the oil only at the top temperature where codistillation will take place. Equally important, the steam must be admitted in such a way as to become saturated with the volatile substances to be removed. Intimate contact is very important and commercial deodoriser manufactures all have their preferred method for getting the best possible contact. Stripping steam amounting to 3–8 % of the weight of oil processed is commonly used.

D. GENERAL DEODORISING EQUIPMENT

1. Batch deodorisers

The first deodorisers were of the batch type and were essentially vertical cylindrical vessels of welded construction with dished or conical ends and height 2–3 times the diameter. Headspace roughly equal to the depth of oil was left at the top to minimise carry-over of oil droplets into the vapour outlet during the violent splashing accompanying deodorisation. Entrainment separators also were used to reduce oil loss. Internal pipe coils served to heat and cool the oil. Stripping steam is admitted at the bottom of the vessel through a perforated pipe arranged to obtain maximum contact of steam with the oil. The vessel should be well insulated, especially the upper half, to minimise condensation and return of distilled volatiles. A three-stage steam ejector will produce the required vacuum. Capacities are typically 10–15 short tons of oil per batch.

Batch deodorisers are now used mainly for low volume runs. Continuous deodorisers were introduced in the 1930s and semi-continuous in 1948. The latter are widely used because they provide many of the advantages of the continuous type while still allowing for easy changes in feedstock. This is important, since unwanted mixing of oils poses serious quality problems, and long runs on a single feedstock are often impractical. It should be noted, however, that several manufacturers of continuous deodorisers now claim to have equipment that does minimise mixing of different feedstocks.

Batch deodorisers are not energy efficient. First the oil must be heated to deodorising temperature with steam or thermal fluid and then it must be cooled with water before being allowed to be stored or shipped. This results in periodic heavy heating and cooling loads. Meanwhile, the vessel is tied up for a long time (8–10 hr). With continuous processing, the entering feedstock, after deaeration, can be heated by passing through coils in the heat recovery tank where hot freshly deodorised oil preheats it, saving about 50 % of the heat required for deodorisation. At the same time, the deodorised oil is cooled significantly without the use of any water. This is a major accomplishment in times of high and rapidly escalating energy costs.

2. Continuous deodorisers

In one double shell type (Fig. 15.1), oil flows downward through a series of compartments mounted within a vertical cylindrical shell with an annular space between them and the shell. Feedstock at 49–54°C (120–130°F) is deaerated by pumping it through an evacuated vessel outside the deodoriser. It is then pumped to the top of the deodoriser for further deaeration and heating. From here, it flows by gravity down through the deodoriser. Normally, in the second uppermost tank superheated steam, Dowtherm, or other heat transfer media is utilised in heating coils to bring the oil up to a deodorisation temperature of 204–274°C (400–525°F). In this and all other tanks, sparge steam is used for agitation. In the next compartment the oil flows over a series of stripping trays countercurrent to the flow of stripping steam injected into the bottom of the vessel.

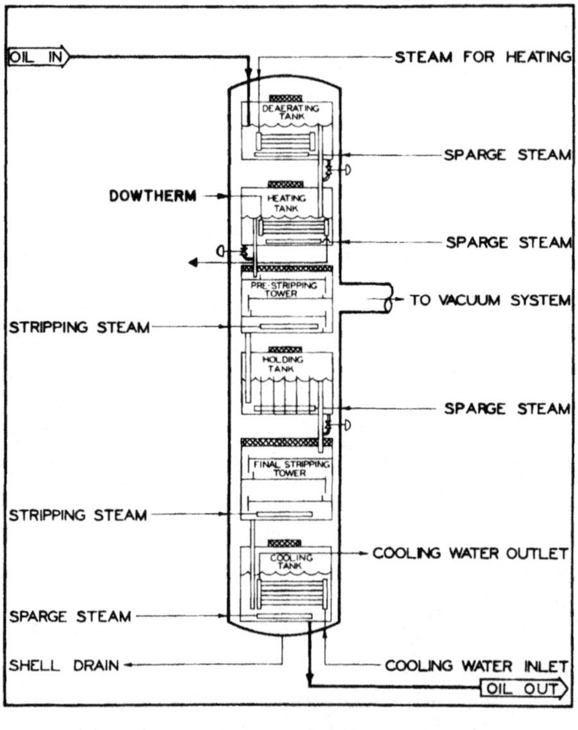

FIG. 15.1. Schematic diagram of a double shell deodoriser. (Courtesy of EMI Corp., Des Plaines, Illinois.)

Oil then enters a holding section to allow time for further heat treatment, including heat bleaching. It next flows to a final stripping section (similar to the third compartment) where the last remaining volatile materials released during the holding period are removed. The oil then enters a cooling section or heat recovery section depending upon whether or not the hot oil is used to preheat the incoming feedstock. Finally, the oil is cooled (with water in coils) to 66°C (150°F) or lower.

Sparge stripping steam and distillate leaving each of the internal sections flow through the annular space surrounding the sections and out into the vacuum system. Any condensate on the inner wall of the outer shell gravitates to the bottom for periodic removal as shell drain condensate.

A less expensive single shell deodoriser (Fig. 15.2) is similar except that a carbon steel vapour pipe connects to the tower in such a way as to allow vapour to pass directly into the vapour take-off pipe from each section. The manufacturer states that for capacities under 15,000 lb per hour the single shell deodoriser is cheaper to fabricate. Both types can be instrumented for automatic feedstock changing.

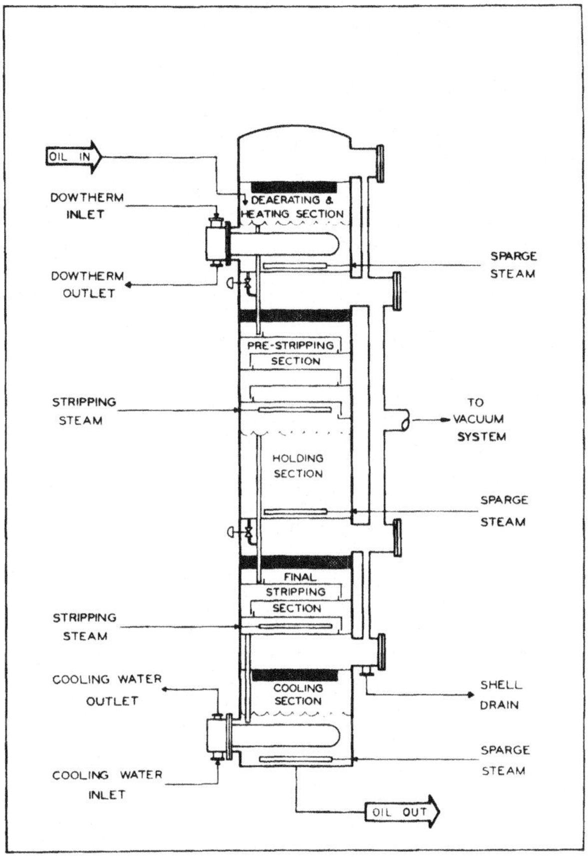

FIG. 15.2. Schematic diagram of a single shell deodoriser. (Courtesy of EMI Corp., Des Plaines, Illinois.)

3. Semi-continuous deodorisers

Based on research by Dr. A. E. Bailey, a semi-continuous unit was designed consisting of a series of trays or tanks mounted within a vertical cylindrical tower (Fig. 15.3). In this equipment, oil moves batchwise downward from one tray to the next for deaeration, heating, deodorisation, and cooling, each conducted in a specific tray. This permits frequent changes of feedstock with very little mixing of successive charges. Four trays are commonly utilised as follows, with 15 min retention in each tray:

1. Deaeration and preheating.
2. Heating to deodorisation temperature.

3. Deodorising tray, using specially designed steam jets for intimate contact.
4. Cooling—by heat exchange with entering oil and with water, or by water alone. Interchange with incoming feedstock can effect a 50% energy saving.

From no. 4, oil is discharged into a built-in drop tank at the bottom of the deodoriser shell from where it is pumped through a polishing filter for further processing or cooled more for storage. Hydrogenated products being more resistant to oxidation may leave the deodoriser as high as 66°C (150°F) without significant flavour damage. With liquid oils a temperature of 38–49°C (100–120°F) is generally preferred to ensure good flavour quality.

Capacities range from about 1100–14,000 kg/hr (2500–30,000 lb/hr).

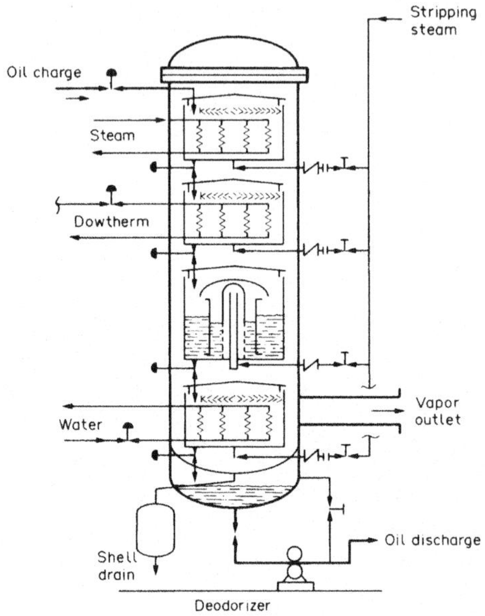

FIG. 15.3. Schematic diagram of a semi-continuous deodoriser. (Courtesy of Chemetron Process Equipment Inc., Louisville, Kentucky.)

E. IMPORTANT FACTORS IN DEODORISATION

1. Air contact

All oil must be thoroughly deaerated before being heated to high temperatures or the oxygen present will seriously affect the quality of the finished oil as a result of the oxidation that occurs. Also, equipment must be carefully monitored for air leakage to avoid characteristic serious off-flavours in the finished oil. In addition, the finished oil must not be allowed to contact the atmosphere until it has cooled sufficiently.

It is difficult to detect small deodoriser leaks because vacuum systems are ordinarily designed to handle more than the normally expected amount of non-condensibles. Thus, the fact that pressure is 6 mm, for example, does not guarantee the absence of leakage. Testing for leaks is especially difficult in commercial deodorisers, unlike glass laboratory apparatus, where the limited air capacity of the vacuum pump usually makes it difficult to obtain the desired vacuum (1 mm Hg or less) if any significant leakage exists. Among the tests used for leaks in commercial equipment are:

(a) *Vacuum drop.* When the oil inlet and discharge pipeline valves are closed and the vacuum system is shut off, the vacuum should not drop rapidly unless there is a leakage. This is a fairly inexact test.

(b) *Soap test.* Another test for larger leaks is run by painting all possible leakage points with soap solution while the system is under 20 psig air pressure.

(c) *Ammonia–sulphur dioxide test.* After sealing off the system, admit ammonia gas to about 5 psig and use compressed air to bring the pressure up to 20 psig. When sulphur dioxide gas is directed around possible points of leakage, white fumes will indicate leakage points.

(d) *Freon test.* After sealing off the system, admit Freon (trademark for dichlorodifluoromethane) until 5 psig is reached, and then compressed air until the pressure is 20 psig. Possible leakage spots are then tested with an electronic Freon detector for high sensitivity or by a propane torch (for fairly small leaks). The flame colour changes to green in the presence of escaping Freon.

2. Stripping steam

An adequate amount (about 3 % of the weight of feedstock) of stripping steam must enter the deodorisation section. Total sparge steam is not an adequate measure.

3. Additives

Citric acid, about 0.01 %, should be added during the cooling stage to chelate metals and increase stability. Antioxidants are optional at this point or may be added in blending later, but it must be remembered that they are less effective in the presence of pro-oxidant metals like iron and copper. Hence, there is a need for chelating agents like citric acid regardless of antioxidant addition.

4. Temperature

The temperature must be checked periodically for constancy and to avoid measurement errors. Panel board readings are not necessarily accurate.

5. Vacuum

Vacuum gauges can and do get out of order. Vacuum needs to be checked periodically with a McLeod or similar vacuum gauge. Also, gauge ranges should adequately cover the anticipated pressures (1–10 mm) accurately. The measurement of any higher pressures, if needed, can be accomplished with portable manometers of wider range.

6. Shut-downs

Air should never be allowed to enter a hot deodoriser or a film of oxidised oil may impart off-flavours to the next oil through. Vacuum should be maintained on short shut-downs, as on week-ends.

7. Nitrogen blanketing

In some installations the deodoriser vacuum is broken with nitrogen, thus saturating the oil with inert gas for stability. In others, the finished oil is pumped into the bottom of a storage tank filled with nitrogen for the same purpose. In general, deodorised oil should be stored for as short a time as possible to minimise oxidation and loss of flavour quality. This is especially true for polyunsaturated oils like soybean.

REFERENCES

1. S. PATTON, I. J. BARNES, and L. E. EVANS, *J. Amer. Oil Chemists' Soc.* 1959, **36,** 280.
2. *National Soybean Processors Association Yearbook and Trading Rules*, Washington, D.C., 1978–79.
3. F. E. SULLIVAN, *J. Amer. Oil Chemists' Soc.*, 1976, **53,** 358.
4. A. M. GAVIN, *J. Amer. Oil Chemists' Soc.*, 1978, **55,** 783.
5. H. A. MOSER, P. C. COONEY, C. D. EVANS, and J. C. COWAN, *J. Amer. Oil Chemists' Soc.* 1966, **43,** 632.

GENERAL REFERENCES

ANDERSON, A. J., *Refining of Oils and Fats for Edible Purposes* (ed. P. N. Williams), chap. 2, p. 155. Pergamon Press/MacMillan, New York, 1962.

Bailey's Industrial Oil and Fat Products, 3rd edition, 1964, chap. 18, p. 897 (4th edition in press), John Wiley & Sons, New York, N.Y.

GAVIN, A. M., "Deodorization and Finished Oil Handling", *J. Amer. Oil Chemists' Soc.* 1981, **58,** 175.

GAVIN, A. M. and R. W. BERGER, "Deodorizer System Modification for Heat Recovery and Steam Refining of Palm Oil", *J. Amer. Oil Chemists' Soc.* 1973, **50,** 466A.

ZEHNDER, C. T., "Deodorization 1975", *J. Amer. Oil Chemists' Soc.* 1976, **53,** 364.

CHAPTER 16

Fractionation and Winterisation of Edible Fats and Oils

Oil and fats contain mixtures of triglycerides of varying melting point and solubility. If cooled carefully, the more saturated higher melting glycerides will solidify first and can be separated. The separation is normally not sharp because the glycerides do not differ sufficiently from each other in their solubility in the melt. Glycerides tend to assume the average characteristics of their component fatty acids, Separation (fractionation) is most satisfactory on free fatty acids or their monoesters, since these fats show greater differences in their physical properties than do glycerides with two or three different acyl groups. For years, chemists in the laboratory have used fractionation, usually in solvent medium, to remove extraneous materials and purify organic compounds, as, for example, synthetic glycerides. Repeated crystallisation is still a standard laboratory technique in organic chemistry.

"Winterisation" is one term used to describe the process by which higher melting glycerides are removed from oils. This process occurred naturally when cottonseed oil was allowed to stand in outside tanks in the southern United States. During the winter, the more saturated glycerides settled leaving a "winterised" oil which would not deposit further solids at refrigerated temperatures (4–10°C, 40–50°F) common in the United States. A winterised salad oil by definition must remain clear for 5.5 hr when placed in a bath of melting ice at 0°C (32°F) in a standard cold test specified by the American Oil Chemists' Society. Currently, only mechanical refrigeration is used in winterising.

Winterisation is now used not only on natural fats and oils but also on processed oils, such as partially hydrogenated soybean oil, to produce a partially hydrogenated liquid oil (PHL oil) suitable for salad oil and salad dressings because of its increased flavour stability.

A. PRINCIPLES OF FRACTIONATION

The efficiency of the separation of liquid and solid glyceride fractions depends heavily on the method of cooling. Rapid cooling causes heavy supersaturation and a large number of small crystals. The result is an amorphous, microcrystalline soft precipitate difficult to filter. This form also tends to change slowly into a metastable α form with characteristics of microcrystallinity and a tendency to develop mixed crystals. On the other hand, slow cooling of the supersaturated oil yields stable β and β' macrocrystals, relatively easy to separate by filtration. Filtration is always a slow and somewhat ticklish operation, however, because the filter cake is easily compressed to a form relatively impermeable to oil.

B. FRACTIONATION PROCESSES

1. Dry fractionation

The process is based on cooling under controlled conditions without the addition of any chemicals, followed by filtration. An early example is the winterisation of cottonseed oil. Here, oil (I.V. about 108) at 21–27°C (70–80°F) is pumped to chillers where it is cooled to 13°C (55°F) in 6–12 hr. First crystals now appear and the cooling rate is slowed so that 12–18 hr are used to drop the temperature to 7°C (45°F). At this point the rate of crystallisation is sufficiently rapid to increase the oil temperature slightly (1–2°C, 2–4°F) even though refrigeration is unchanged. The temperature then drops as before; and when it has dropped slightly below the previous minimum, cooling is discontinued and the batch held at this temperature for a considerable time. Generally after 12 hr the filtered oil will have a cold test of about 20 hr. Filtration may be carried out in ordinary plate and frame presses or other suitable filters of large capacity. The yield of winterised oil will be about 75–80% with an iodine number of 110–114.

A current example of dry fractionation is the Tirtiaux process.

2. Lanza fractionation

Lanza fractionation is similar to the dry process as far as the separation stage. Then an aqueous detergent solution is added, which replaces the oil phase on the surface of the crystals and allows the suspension formed to be separated by centrifugation. The fat crystals are melted in a heat exchanger and the oil–water mix separated by centrifugation. Typical of this process is the Alfa-Laval Lipofrac system.

3. Solvent fractionation

Cooling of a fat diluted with solvent usually results in a better fractionation on cooling. Acetone, hexane, and 2-nitro propane are typical of solvents used. Acetone has been used to produce cocoa butter substitutes from hydrogenated liquid oils, shea fat, and palm oil. A Bernardini process, using hexane, has been employed for palm oil fractionation.

C. EXAMPLES OF FRACTIONATION

1. PHL oil

The term PHL oil refers to partially hydrogenated liquid soybean oil (also called "HWSBO", hydrogenated winterised soybean oil). It is a salad oil competing with winterised cottonseed oil in the United States. Soybean oil alone will withstand salad oil refrigerator temperatures of 4.5–10°C (40–50°F) without solidifying or clouding; but where it is to be used as a dual purpose salad-cooking oil, partial hydrogenation is usually employed to improve odour, flavour, and oxidation stability during use at high temperatures. This hydrogenation saturates double bonds and also produces some *trans* acids, causing the oil to cloud or show precipitate at low temperatures. To avoid this, the soybean oil is usually hydrogenated selectively to an I.V. of 110–115 from the original I.V. of 130–135 before winterising. Below I.V.s of about 105 the loss in yield does not usually compensate for any improvement in flavour or stability.

During the winterisation process, slow cooling and agitation are essential to avoid producing small crystals that make filtration difficult or impossible. Plate and frame presses have been the standard of the industry, but pressure leaf filters and continuous vacuum filters are being used more at the present time. Centrifugation is an obvious choice if it can be done; but because of the commercial advantages, processors keep this information proprietary. Yields of PHL oil range from 75 to 85 %. Comparison of the fatty acid composition of soybean oil and several PHL oils is shown in Table 16.1.

TABLE 16.1. *Comparative Fatty Acid Composition (%) of PHL and Soybean Oils*

Fatty acid	Soybean oil	PHL oils 1	2	3	4
Palmitic 16:0	9.2	10.0	9.9	8.4	10.2
Stearic 18:0	4.7	6.1	6.2	5.3	5.7
Oleic 18:1	24.0	38.4	38.8	48.0	32.2
Linoleic 18:2	54.2	41.3	40.6	34.8	37.0
Linolenic 18:3	8.0	4.1	4.4	3.5	3.9
I.V.	135.4	116.3	115.2	110.7	111.5

2. Fractionated palm oil

Palm oil is quite stable to oxidation, but is semi-solid at ordinary temperatures, 21–27°C (70–80°F). Since a liquid stable oil is quite attractive, several plants have begun to fractionate palm oil to produce a liquid oil (65–70 %) (palm olein), melting point around 18–20°C (64–68°F) and a stearin fraction (30–35 %) of m.p. 48–50°C (118–122°F). For use in some tropical countries the liquid palm oil fraction may be mixed with coconut oil to produce a stable liquid cooking oil, while part of the solid palm oil fraction may be blended with palm and coconut oils for margarine production.

3. Liquid shortening

Plastic shortenings, although satisfactory in performance, suffer some disadvantages in that they are more difficult to handle and measure. Many bakeries prefer a liquid shortening that can be pumped and metered. This also avoids the risk of melting a plastic shortening and having it crystallise on cooling to an unacceptable structure not originally present. Moreover, in fast food restaurants with relatively unskilled personnel, the adding of liquid oil to the fryer is safer and more convenient.

The general principle in making a liquid shortening from soybean oil involves hydrogenation to increase stability, followed by removal of solids through winterisation. In one process soybean oil is hydrogenated to an I.V. of 81–84 and, after removal of the nickel catalyst, cooled slowly from 60°C (140°F) to 15.6°C (60°F). It is claimed that a 48–54 % yield of liquid fraction (I.V. 86–88) is obtained with less than 1% solids at 10°C (50°F) and free of dispersed solids on storage at temperatures over 15.6°C (60°F).

In another process soybean oil is hydrogenated, winterised, and the liquid portion again hydrogenated and winterised to produce a liquid oil with a good stability at deep fat frying temperatures.

4. Winterisation of cottonseed oil and of blends with soybean oil

One problem with any winterisation is to find a use for the stearin fraction. Normally, plants have ample saturated fat and try to avoid making any more. In a process reported by Gooding,[1]* blending soybean oil and cottonseed oil reduces the stearin yield for cottonseed oil (15–25%) and for soybean oil (7–8%) to as little as 4–5% for a mixture. In addition, it is claimed that salad dressing and mayonnaise made with the liquid fraction show improved freeze resistance.

D. CRYSTAL INHIBITORS

A crystal inhibitor is a material which retards crystal formation when added to an oil, thus increasing the so-called "cold test" (usually calling for 5.5 hr at 0°C (32°F) without any clouding or precipitation). Several additives have been developed, but only two are approved for use in the United States by the Food and Drug Administration. In this connection it is worth mentioning that each country has its own regulations and what is approved in one country may not be approved in another. Moreover, some countries are more strict in some respects and more lenient in others. Regulations of the individual country must always be considered in fat and oil processing.

One widely used crystal inhibitor, approved in the United States, is oxystearin made by a controlled heat treatment of hydrogenated soybean oil or cottonseed oil in air and in the presence of a catalyst. This is permitted in cottonseed and soybean cooking and salad oils at a maximum of 0.125%. Polyglycerol esters of stearic, oleic, and coconut fatty acids are also permitted when their use is not prohibited by Food Standards.

For obvious reasons, many buyers do not want a liquid oil containing crystal inhibitors, but prefer one with an acceptable cold test before any additive use.

E. STABILISERS

Although HWSBO is reportedly more stable than soybean oil, the latter is quite suitable in salad dressings and similar products that will be stored and used cold. The advantage of HWSBO comes mainly if heating is to be done. In any case, either oil should be treated with citric acid to chelate metals. Added antioxidants are less important.

In frying fats, added methyl silicones minimise foaming and may help in improving room odour when frying.

* Superscript numbers refer to References at end of Chapter.

REFERENCE

1. C. M. Gooding, U.S. Patents 3,048,491 and 2,627,467.

GENERAL REFERENCE

Weiss, T. J., *Food Oils and Their Uses*, AVI Publishing Co., Westport, Connecticut, 1980.

CHAPTER 17

Interesterification*

A. INTRODUCTION

To prepare fatty acid esters it is not necessary to start with a fatty acid and an alcohol. For example, if a triglyceride (R^1COOR^2) is heated with an alcohol (R^3OH), the following reaction takes place:

$$R^1COOR^2 + R^3OH \xrightleftharpoons{\text{catalyst}} R^1COOR^3 + R^2OH$$

If R^3OH is methanol, methyl esters of the fatty acids are made. If R^3OH is glycerine, an equilibrium mixture of mono- and diglycerides is made. This is called "*alcoholysis*".
If a triglyceride (R^1COOR^2) is reacted with a carboxylic acid (R^3COOH), the reaction is:

$$R^1COOR^2 + R^3COOH \rightleftharpoons R^3COOR^2 + R^1COOH$$

In this way a C_{12} acid can be made to replace C_{16} and C_{18} acids in the triglycerides. This is called "*acidolysis*".

In "*interesterification*" a natural oil or fat has its acyl groups randomly distributed in the glycerides, changing the physical properties depending upon how far the original oil or fat varied from a random distribution initially.

This treatment is important because the properties of a fat or oil depend not only on the fatty acid composition but also on the distribution of the fatty acids in the glycerides. For example, cocoa butter confectionery uses are attributed to the fact that it contains mainly glycerides with an unsaturated acid in position 2 and two saturated acids in positions 1 and 3, *viz.*

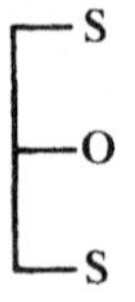

where S is a saturated and O an unsaturated acid. After interesterification there is random distribution of the fatty acids so that only about 12.5 % of the cocoa butter glycerides exist in the original structure and the physical properties are changed completely. Similarly, as mentioned in

*See also Chapter 9.

Chapter 9, interesterifying soybean oil raises the melting point from −7 to +6°C and that of cottonseed oil from 10 to 34°C, despite no change in fatty acid composition.

On a practical basis, interesterification affords a method for changing the existing (non-random) glyceride structure of a fat to that produced by a random distribution of the fatty acids among the glycerides. Also, fatty acids may be added to a fat by interesterifying with another fat. The method can be used effectively to produce tailor-made fats and margarine oils. Since about 1950 interesterification processing has been used to such an extent that in Europe it is said to be employed as frequently as hydrogenation in many areas. In the United States major utilisation has been in the treatment of lard.

B. TYPES OF INTERESTERIFICATION

1. Random

In random interesterification treatment of an oil or fat with an interesterification catalyst converts the mass into glycerides in which the fatty acids present are randomly distributed. For example: An equimolar mixture of saturated (S) and unsaturated (U) fatty acids present only as trisaturated glycerides (GS_3) and triunsaturated glycerides (GU_3) can be interesterified to yield GS_3 (12.5%), GU_3 (12.5%), GS_2U (37.5%), and GUS_2 (37.5%). The amounts of the various glycerides can be calculated as follows:

$$\%\ \text{triglyceride } GS_3 = \frac{S \times S \times S}{10{,}000}$$

$$GU_3 = \frac{U \times U \times U}{10{,}000}$$

$$GSU_2 = 3\left(\frac{S \times U \times U}{10{,}000}\right)$$

$$GS_2U = 3\left(\frac{S \times S \times U}{10{,}000}\right)$$

where S and U are the molar percentages of saturated and unsaturated acids, respectively.

2. Directed

In this type of interesterification some of the fatty acids or their glycerides are removed from the equilibrium reaction, forcing it to continue to a new equilibrium. In one method, low molecular weight fatty acids are distilled out of the mixture. More commonly, the reaction is run at a low temperature such that the higher melting glycerides formed by the reaction crystallise out forcing the reaction to proceed in that direction.

C. CATALYSTS

The most common catalysts are sodium methylate, sodium metal, or a liquid sodium–potassium alloy. It is essential that the fat contain no substances which might destroy the catalyst. This means that the fat must be well refined, dried, and heated to about 150°C (302°F) under an inert atmosphere before the catalyst is added. Analytically, moisture below 0.01 %, free fatty acids below 0.05 %, and low peroxide value are guidelines.

D. PROCEDURE

In random interesterification the fat is heated to about 80°C (176°F) in the presence of a catalyst (0.05–0.20 %) and with good agitation. Reaction time is about ½hr. At the end of the reaction, the catalyst is inactivated with water or dilute acid.

In directed interesterification at low temperatures (26–38°C, 80–100°F), the reaction is much slower, due to the lower temperature; and for this reason, reaction time may be several hours to a day or possibly more. A solvent may also be employed to facilitate crystallisation of the rearranged more saturated glycerides. It is essential that the catalyst be inactivated with water or dilute acid before the fat is recovered in order to avoid disturbance of the equilibrium.

Some loss is always encountered because of the alkaline nature of the catalyst. This normally amounts to about 1 % neutral fat for each additional 0.1 % catalyst used over an optimum of about 0.1 %.

E. EXAMPLES

1. A solid fat containing about 60 % essential fatty acids can be obtained by directed interesterification of sunflower oil and blending with 5 % hard fat.
2. Random interesterification of lard results in an improved shortening by reducing the graininess of the original fat, caused by the presence of large crystals of disaturated glycerides (stearic, palmitic, oleic) in which the palmitic acid is in the 2(middle)-position and the oleic in the 1-position. Random interesterification reduces the amount of these glycerides.
3. A palm oil of m.p. = 51°C containing 30 % S_3 and 26 % U_3 can be made by directed interesterification of palm oil of m.p. = 39°C containing 7 % S_3 and 6 % U_3. Fractionation permits a higher yield of liquid oil than from the untreated oil.
4. Margarine oil made with a high content of lauric oils is typically low in melting point and has a short plastic range. This produces a margarine that is hard in the refrigerator but melts partly at room temperature. The remedy is to remove or decrease the amount of lauric triglycerides. This is done by interesterifying the coconut oil with an oil such as palm and then blending 60 % of the interesterified mixture with 40 % of an oil like sunflower.
5. A high essential fatty acid low *trans* acid margarine can be made by interesterifying a coconut-soy hardstock and blending with soybean oil. The more soybean oil in the interesterification mixture the lower the melting point, assuming the same total soybean oil in the final product.

GENERAL REFERENCES

HUSTEDT, H. H., "Interesterification of Edible Oils", *J. Amer. Oil Chemists' Soc.* 1976, **53,** 390.
SCREENIVASAR, B., "Interesterification of Fats", *J. Amer. Oil Chemists' Soc.* 1978, **56,** 796.

CHAPTER 18

Margarines and Shortenings

Margarine and shortening have much in common. Both are largely or entirely fat, containing one or more hydrogenated or unhydrogenated oils or blends of both types, which have a major influence on properties. Margarine contains an aqueous phase and is a water-in-fat emulsion (commonly 80% fat) in which the water droplets are kept separated by the fat crystals. Because of the similarity in fat structure, margarine and shortening will be discussed together before emphasizing their differences.

A. DEFINITIONS

Margarine was invented by the French chemist, Mége-Mouriés, as a butter substitute in the late 1860s. Originally, oleo oil (thus the term "oleomargarine") was mixed with milk and the mixture digested with the tissue from cow udders to obtain a butter-like flavour. This flavour was later found to be due to the action of bacteria in the milk and the use of mammary tissue was abandoned in favour of controlled fermentation of milk by butter cultures. Margarine compositions vary somewhat by country. In the United States a typical margarine would contain:

Fat—Minimum, 80%.
Water—With or without edible protein.
Emulsifiers—"Safe and suitable". Originally these were only lecithin and mono/diglycerides.
Salt—Sodium or potassium chloride or none.
Preservatives—Sodium benzoate, benzoic acid, and potassium sorbate permitted at specific levels to control microbiological growth, especially in the absence of salt.
Antioxidants—At a safe and suitable level.
Colours—At a safe and suitable level.
Flavours—At a safe and suitable level.
Vitamins—Vitamin A—mandatory (15,000 units)
Vitamin D—optional (2000 units)
Vitamin E—excluded (but often used in Europe).

Early margarine quality left something to be desired, especially in view of many restrictions promulgated by anti-margarine interests. Today margarine is fully comparable to butter in quality.

Initially, the term "shortening" was reserved for lard used to make bread, as well as pies with a tender flaky crust. The term has since expanded to include all commercial fats and oils except oil-derived products, such as margarines and other high fat-content products containing various non-fat materials. The term "shortening" has a literal meaning in that shortenings actually do "shorten" or tenderise baked foods by preventing the cohesion of wheat gluten strands during mixing, thus really shortening them.

B. HISTORY

Man's first fat products were probably rendered from the carcasses of wild animals. Later, the body fat of domestic animals became an important article of commerce. This supply was increased as the whaling and sealing industries provided large quantities of marine oils. In time, lard, the body fat of pigs, came to be the preferred fat for edible purposes. This was partly due to flavour, but mostly because lard had acceptable consistency for easy incorporation into baked goods. Unfortunately, the supply of lard is a by-product of the meat packing industry and the supply is not directly related to demand. As the production of lard fell below the demand for plastic shortening in the lard-consuming countries, the supply of shortening was increased by blending a small amount of oleostearin or other hard animal fat with a relatively large amount of vegetable oil. Later, with the development of hydrogenation, it became possible to omit the animal fat entirely and produce a plastic product from liquid vegetable oil or marine oils. In the United States the rapid growth of the cotton industry resulted in large quantities of available cottonseed oil and shortenings were developed to provide an outlet for the oil. So-called "lard compounds" were made by blending hard and soft fats to provide a lard substitute. At first, this was done by meat packers, but as technology developed animal fat could be replaced by suitably processed vegetable oil without difficulty. Today, plastic shortening can be produced with a higher content of polyunsaturated fatty acids (PUFA) than before; as high as 22–35% as compared to 6–15% in 1961. This has been done in response to nutritional results which indicate that a greater consumption of these acids is desirable for good health.

C. STRUCTURE OF A PLASTIC FAT

Shortening at first glance appears to be a soft but more or less homogeneous solid. Careful inspection under magnification shows it to consist of a mass of very small crystals in which a considerable amount of liquid oil is enmeshed. Further study shows that the crystals do not form a continuous structure but that each behaves as a discrete particle, capable under the proper shearing stresses of moving independently of other crystals. This is a characteristic of a plastic solid, the distinguishing feature of which is to behave as a solid and resist small stresses, but to be easily malleable when the deforming stresses exceed a certain minimum value. Thus, a firm plastic solid will not deform under its own weight, but can be shaped as desired under proper stress.

In general, to be plastic a material must:

(1) Consist of two phases, capable of acting as solid and liquid.

(2) Have a solid phase sufficiently finely dispersed so that the entire mass is held together by cohesive forces.

(3) Have a proper ratio of the two phases, i.e. the solid particles must not be so predominant as to form a rigidly interlocked structure, nor so few that the mass can flow without the particles blocking it by forming jams or arches.

A viscous material, in contrast to a plastic material, will deform somewhat at the slightest stress and so viscosity is a measure of its consistency. A plastic, on the other hand, will absorb limited stresses elastically ("yield value") before it becomes permanently reshaped. The higher the "yield value", the greater the stress that can be accepted before the material is deformed. In general, high yield values and high viscosities go hand in hand.

D. PRODUCTION OF SHORTENING

Plastic shortenings are normally made from one or more partially hydrogenated vegetable oil base stocks or mixtures of such stocks with animal fat base stocks. Hard fats are added to extend the plastic range. Typical plastic shortenings have a relatively flat SFI(Solid Fat Index) curve over a temperature range of 15–32°C (60–90°F) with solids ranging from 15 to 30%. In one formulation, 10% of hard oil would be used with 90% of a partially hydrogenated base oil in the iodine value range of 65–80. Alternatively, two partially hydrogenated base oils might be used. It is essential that the hard fat crystallise in the proper form, β'. This is because β' hard fats cause the entire fat to crystallise in that form, producing a shortening crystallising in small needles, with a smooth texture and having aeration and creaming properties excellent for cakes and icings. If a shortening crystallises in the β form, it tends to be grainy; with large crystals, it is difficult to aerate, but is excellent for pie crusts (like lard in this respect). Table 18.1 lists hard stocks with β and β' crystallising tendencies.

TABLE 18.1. *Fats and Oils Classified as to Crystal Habit*

β type	β' type
Canbra[a]	Cottonseed
Cocoa Butter	Herring
Coconut	Lard (Modified)[b]
Corn	Menhaden
Lard	Milk Fat
Olive	Palm
Palm Kernel	Rapeseed
Peanut	Tallow
Safflower	Whale
Sesame	
Soybean	
Sunflower	

[a] A low erucic acid rapeseed oil.
[b] Wholly or partly interesterified.

Shortening production involves: (1) preparation of the individual base stocks and hard fats, (2) formulation, (3) solidifying and plasticising the blend, (4) packaging, and (5) tempering, if needed. Item (1) has been discussed under hydrogenation. Formulation consists merely in mixing the ingredients in preparation for step (3). Solidification and plasticising are generally accomplished using scraped surface heat exchangers (Fig. 18.1), such as the Votator "A" unit. In this equipment a steel shaft rotates in a tube which is cooled externally by ammonia. Scraper blades on the rotating shaft, moving at high speeds, are pressed against the cooled inner surface by centrifugal force. The high internal pressure and shearing action cause fast nucleation and crystallisation during the short residence time of a few seconds. Typically, the melted fat blend, plus optional ingredients, is

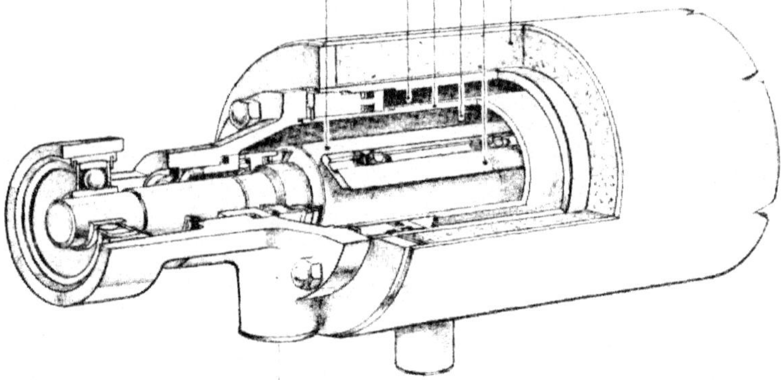

FIG. 18.1. Diagrammatic sketch of a votator scraped surface heat exchanger. (Courtesy of Chemetron-Process Equipment, Inc., Louisville, Kentucky.)

chilled rapidly from 46–49°C (115–120°F) to 16–18°C (60–65°F) in one or more "A" units. The supercooled melt is then pumped to a "B" unit, a large diameter tube fitted with stator pins on the cylinder walls and a high speed rotating shaft equipped with rotor pins. This mechanically works the fat as it passes through the unit, the crystals growing and the mass partially solidifying. The "working" extends the plastic range of the product. Inert gas (10–20%) or air may be added here and dispersed in the shortening as small bubbles to improve the whiteness of the product. This may also contribute to creaming ability. The plastic mass is now pumped through a homogenising valve and finally to package fillers. Containers may range in size from 0.45 kg (1 lb) to 172 kg (380 lb) drums. When tempering is needed, the packaged product is normally held for 24–72 hr in a constant temperature room, usually at 27–32°C (80–90°F). During tempering, crystallisation proceeds slowly and the crystal structure is stabilised against changes that might otherwise take place during subsequent temperature variations encountered during normal handling and storage. However, should the product be allowed to be warmed above the melting point of the lowest melting polymorphic form, the structure is lost and the product must be melted and reprocessed.

E. TYPES OF SHORTENING

1. Plastic shortenings

These are used in their semi-solid state. They may contain as optional ingredients emulsifiers, antioxidants, metal scavengers, pigments, and flavours. Emulsifiers are important for texture control in bread and cakes, aeration in cakes, toppings, etc.; dough conditioning and anti-staling in bread, and wetting agents for coffee whiteners and instant foods. Besides mono- and diglycerides, emulsifiers include succinylated monoglycerides, ethoxylated mono- and diglycerides, polysorbate 60, calcium stearoyl-2-lactylate* and sodium stearoyl-2-lactylate.* In recent years, large volume users have tended to avoid the handling problems connected with shortening in containers and have preferred receiving their shortenings in bulk form, as a liquid melt, in tank trucks or railroad tank cars. On receipt the liquid is normally held at 3–6°C (5–10°F) above its melting point (closed capillary method) until needed. It is then chilled and plasticised at that plant. According to Bauerlen[1]† bulk handling is more typically used for frying and baking shortenings that can be used in the liquid state without crystallisation.

2. Pourable shortenings

Pourable shortenings include fluid suspensions, fluid emulsions, and clear liquids. Fluid suspensions are opaque, not clear, and consist of a base to which has been added suspended solids in the form of hard fats or emulsifiers. Unlike plastic shortenings, a β crystal form is desired in both the hard fats and emulsifiers. In one scheme a rapidly chilled mixture of base stock and hard fat is held in a tank until it forms a soft solid, after which it is stirred gently to fluidise before packaging. Fluid emulsions are a hydrated form of fluid suspensions.

Clear liquids include cooking oils, salad oils, and the liquid components resulting from the fractionation of partially hydrogenated oils. These are often not emulsified for technological reasons, although this is possible with certain emulsifiers.

The development of continuous bread baking is responsible for much of the demand for liquid or fluid shortenings that can be pumped, metered rather than weighed, and conveniently handled and stored in volume. Actually, with advances in emulsifier technology, large bakeries are switching from animal fats and plastic shortenings to refined and deodorised vegetable oils used together with emulsifiers. Soybean oil, with emulsifiers, is now being used by the wholesale bread baking industry in the United States.

3. Solid (dry) shortenings

Powdered or pelletised shortening used for baking mixes can be prepared in a number of ways. In general, the solid shortenings are sprayed into cold air and collected as a powder or solidified on a "chill roll", cooled further and ground. Alternately, a prepared, gelatinised, powdered starch may be used to absorb the shortening or oil.

F. SHORTENING USES

1. Baked goods

Included here are chemically leavened baked goods, such as cookies (biscuits), cakes, cake doughnuts, as well as Danish and puff pastries. In addition, yeast raised baked goods form a group

* Salts of $C_{17}H_{35}COOCH(CH_3)COOCH(CH_3)COOH$.
† Superscript numbers refer to References at end of Chapter.

that encompasses sweet rolls, bread, rolls, buns, coffee cakes, and yeast-raised doughnuts. In the United States household shortenings for baking "from scratch" are losing ground to prepared mixes and refrigerated doughs which are more convenient, less time-consuming, but of perfectly acceptable quality.

2. Fried foods

For frying, fats serve both as a heat transfer medium and as a source of flavour and nutrition since some of the fat is absorbed by the food being fried. In household use where the fat is not kept hot for a prolonged time and is often discarded after limited use, shortening requirements are not strict and even liquid polyunsaturated oils, like safflower, are sometimes preferred. In commercial establishments, however, it is important to have good oxidative stability, high smoke point, minimum foaming, minimum colour darkening, and good heat transfer. Despite extensive research there is still no one method for chemically evaluating frying fat quality. An important factor is "turnover", the rate at which the fat absorbed by the food is replaced by fresh fat. Ideally, turnover should be as high as possible, economics considered. In some installations, fat absorption by the product fried is so large that make-up fat is about all that is needed. In small volume restaurants, on the other hand, it may be necessary to replace all the fat at rather frequent intervals to avoid foaming, darkening, greasiness, and off-flavours. In general, frying fats should be heated to as low a temperature and for as short a time as possible, kept away from air, protected from contamination by solid food particles and by oils extracted from such foods as chicken. Free fatty acid measurement parallels fat breakdown in some cases but, unfortunately, not in all.

3. Icings and cream fillers

These products have specialised requirements for consistency, flavour, emulsification, and shelf life, depending on their usage.

4. Frozen foods

The same requirements apply here as is common to foods not frozen since much of the food is precooked and needs only defrosting and heating.

G. PRODUCTION OF MARGARINE

Since margarine is a water-in-fat emulsion, we can discuss the two phases separately.

1. Fat phase

The objective is to produce a fat mixture with a fairly steep SFI curve so that the product is firm in the refrigerator but spreads rather easily on removal and melts rapidly in the mouth. It should also crystallise as β' crystals. In United States common practice, selectively hydrogenated oil components are blended for this purpose. Depending upon oil costs and availability, however, treatments, such as corandomisation, fractionation, interesterification, and direct blending of natural fats may be preferred. In general, the more saturated glycerides provide structure with increasing unsaturation providing lubricity and increased nutritional value. Wiederman[2] has described four ways in which components can be blended:

(a)	High amount High I.V. oil	+	Low amount Low I.V. fat
(b)	Intermediate level High I.V. oil	+	Intermediate level Intermediate I.V. fat
(c)	Low amount High I.V. oil	+	High amount Intermediate I.V. fat
(d)	Blend of intermediate I.V. fats		

Blend (a) finds use in fluid margarines, high liquid oil stick products, and some highly polyunsaturated soft tub products; (b) for high polyunsaturated stick and soft tub products; (c) for low polyunsaturated soft tub products; and (d) for all hydrogenated stick products.

The fat phase will also contain the oil-soluble ingredients, such as mono- and diglycerides and lecithin, colouring matter, and vitamins.

2. Aqueous phase

This phase comprises up to 20% of normal margarine. Traditionally this was all cow's milk, but may now be all water, with or without some edible protein component. In current practice for the production of the aqueous phase, a "milk" may be prepared by adding dried protein to water and pasteurising. Next, the "milk" is pumped to a milk ripener in which it is inoculated with about 1% of a starter culture for the production of lactic acid, diacetyl, and other flavours. When the proper pH is reached or the required amount of flavour has developed, the water-soluble ingredients (salt and preservatives) are added and the mass kept at 5–8°C (41–46°F) until ready for use. "Ripening" is not needed if adequate flavour concentrates are used.

3. Blending and chilling

In batch processing, the fat and aqueous phases are scaled into an emulsion mix tank or "churn" where they are emulsified under high-speed agitation. In continuous processing automatic proportioning equipment is used instead. The emulsion is now pumped to an emulsion hold tank feeding a Votator "A" unit and then through a "B" unit, either quiescent or working, before printing or filling. Depending upon the product, various arrangements of "A" and "B" units may be used.

4. Tempering

Tempering usually involves holding at an elevated temperature (25–35°C, 77–95°F) for 2–3 days based on past experience. Tempering improves plasticity, creaming properties, and baking performance. The slow crystallisation which occurs favours the crystal growth, which results in improved creaming, i.e. uptake of water and air in batters. Table and kitchen margarines are not ordinarily tempered.

H. MARGARINE USES

1. Table

At least in the United States most margarine is used as a table spread in place of or as an adjunct to butter. Since easy spreadability at refrigerator temperatures is highly desirable, manufacturers have developed soft stick, whipped, and various "tub" products—all designed to make spreading easier. Solid Fat Indexes, common in some of these products, are shown in Table 18.2. In addition to spreadability, nutrition-conscious consumers are demanding less saturation and more and more polyunsaturated content. This is illustrated in Table 18.3, taken from recent literature. Furthermore, to satisfy calorie conscious consumers, there are now available lower fat content "vegetable oil spreads" in the 40–60 % fat content range. To date, preferred products appear to be mostly in the 60 % fat range. There are also some "liquid" or "pourable" margarines of normal fat content, but easily dispensable for use on potatoes, vegetables, as well as for a fast easy bread spread. These are also being adopted for industrial use in prepared and frozen foods.

TABLE 18.2. *Typical SFI Values for United States Margarines*[2] (2)

	10°C 50°F	21.1°C 70°F	26.7°C 80°F	33.3°C 92°F	37.8°C 100°F
Stick	28	16	12	2–3	0
80 % liquid oil print	15	11	9	5	2
Soft tub	13	8	6	2	0
Liquid oil + 5 % hard fat	7	6	6	5.4	4.8
Bakers	27	18	16	12	8
Roll-in	29	24	22	16	12
Puff-past	26	24/21	23/20	22/16	21/15

TABLE 18.3. *Margarine Compositions*[3]

	Fatty acids (g/100g Margarine)			P/S ratio*
	Saturated	Mono-unsaturated	Poly-unsaturated	
Stick or brick	13.1–22.9	22.3–53.8	9.2–41.2	0.7–3.1
	av. = 15.8	av. = 37.4	av. = 22.7	1.4
Soft tub	9.3–17.4	16.1–42.8	21.4–48.4	1.5–4.7
	av. = 15.0	av. = 31.2	av. = 30.1	2.0
Liquid	13.1	22.0	40.7	3.1

* Ratio of polyunsaturated to saturated fatty acids.

2. Bakery

These products are normally specially processed for bakery use, although a considerable amount of the table-type is also used. They are usually modified somewhat by the addition of 4–8 % hard fat as a plasticiser and are used for biscuits, pound cakes, and pastry. "Roll-in" margarines, used for Danish pastry, are higher melting and contain higher Solid Fat Indexes at 33.3°C (92°F) and 37.8°C (100°F). "Puff pastry" normally contains about 90 % fat and is usually churned with water rather than milk or other protein-containing phase. It is used for turnovers, pastry shells, Napoleons, and similar products.

REFERENCES

1. R. J. Bauerlen, *Bakers' Digest* **47**(5). 91, 130.
2. L. H. Wiedermann, *J. Amer. Oil Chemists' Soc.* 1978, **55,** 823.
3. J. L. Weihrauch, C. A. Brignoli, J. B. Reeves III, and J. L. Iverson, *Food Technology*, Feb. 1977, 80.

CHAPTER 19

Flavour Stability and Antioxidants*

A. INTRODUCTION

Off-flavours and odours in oils and fats are caused by the reaction of oxygen with the double bonds of unsaturated fatty acids. The theory is discussed in Chapter 7. There have been many attempts to predict stability through various chemical tests, but nothing is as satisfactory as trial under the conditions to which the oil or fat is to be exposed. Other tests are of limited usefulness. In general, oxidation may be prevented or hindered by:

Avoiding contact with oxygen.

Reducing unsaturation in the product.

Avoiding conditions favouring oxidation, namely:

Exposure to light.

Elevated temperatures.

Presence of pro-oxidants.

B. ANALYTICAL METHODS OF STUDYING FLAVOUR STABILITY

1. Peroxide value (P.V.)

This is expressed in milliequivalents of peroxide per kilogram of fat. Fresh fat should have a low P.V. A specification of "1" or below is not uncommon for finished oil. The presumption is that an elevated P.V. indicates that oxidation has begun, that the fat is in or past the induction period and oxidation may proceed quickly from now on.

2. Schaal test

This is expressed as the time required for a loosely covered sample held at 60°C (140°F) to develop a rancid odour or flavour. Sometimes a definite P.V. is used as the end point.

* See also Chapter 7.

3. Active oxygen method (A.O.M.) (Swift stability test)

This has been one of the most widely used methods for determining fat stability since it was developed in 1933. It is based on bubbling washed air through a fat sample held at 97.8°C (209.0°F) until a predetermined P.V. is obtained. For animal fats a figure of 20 is common; 100 is common for vegetable fats. The time is expressed in hours. Many specifications require a minimum A.O.M. value. More recently it has become common to use an 8-hr A.O.M. maximum, that is, the P.V. that is developed after 8 hr of treatment. An A.O.M. run at 110°C (230°F) has also been used to speed up the process and decrease analytical time. In any event, it must be kept in mind that the A.O.M. merely demonstrates what the fat will do under the test conditions. It does not correlate well in many cases with shelf-life, particularly in the case of salad oil. Very high figures, such as 500 hr, may greatly overstate the stability of the fat under conditions of use.

Alternative accelerated procedures developed and used in Europe depend on some change of property during storage of fatty sample in an atmosphere of oxygen at an elevated temperature. The Fira-Astell apparatus measures changes in oxygen pressure and the Rancimat measures volatile short-chain acids.

4. Anisidine value

This method is used by some for a measure of secondary oxidation products, high molecular weight carbonyl compounds, indicative of prior oxidation. Thus, an oxidatively abused fat could show a low P.V. after deodorisation, but presumably the anisidine value (little changed by processing) would warn of this past history. The method is seldom used in the United States, perhaps because most oils are of domestic origin and relatively fresh.

5. TOTOX value

This is expressed as the sum of the Anisidine Value plus 2 times the P.V. It appears to be used mostly by researchers, primarily in Europe.

6. Volatiles

This involves a G.L.C. measurement of very small amounts of degradation products arising from oxidation. Still mostly a research tool, it gives clues as to whether the sample was exposed to light or heat alone. Also, volatiles have been correlated with organoleptic flavour evaluation by several laboratories and could possibly replace organoleptic evaluation sometime in the future. It is also a good measure of efficient deodorisation. For example, a well-deodorised soybean oil will normally run under 5 ppm in volatiles. Some investigators even hope that volatiles may be used as a predictive tool as further data are obtained.

C. EVALUATION OF SHELF-LIFE

Ideally this should be done under the conditions to be found in proposed usage, but is usually modified for speed of results and convenience. After a specified storage the flavour of the oil is evaluated organoleptically as described later.

1. Storage

Freshly deodorised oils are usually stored in stoppered bottles in the light and in the dark. The air volumes above the oil ("head space") must be kept constant for consistent results. This storage is intended to simulate conditions in retail stores which are dark at night and illuminated, often by fluorescent lights, during the day. Plastic shortenings, of course, would have light striking only the outer surface. Figure 19.1 shows the flavour scores of a typical soybean oil as purchased, after being stored in the light. P.V.s are usually determined along with flavour scores.

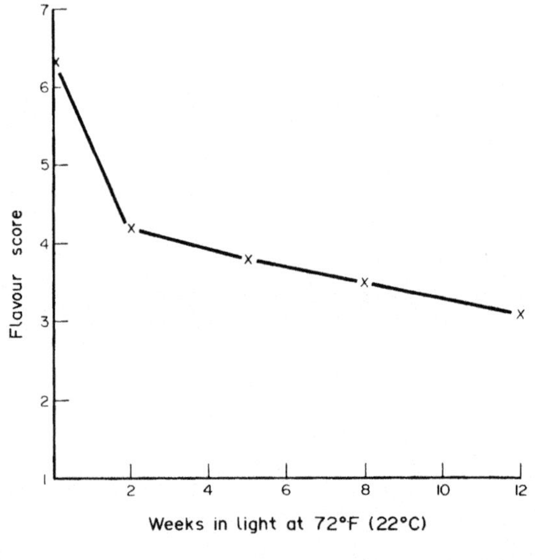

FIG. 19.1.

2. Accelerated storage

Increasing the temperature is an obvious way to speed up oxidation, although it may be misleading if the temperatures used are too different from those the product will be exposed to. Another method suitable for storage in the light is to increase the illumination or to use ultraviolet light which is much more deleterious than artificial light. A "light box" developed by the Northern Research Center of the United States Department of Agriculture rotates bottles containing liquid oil in a container whose sides are equipped to illuminate the oil under specified conditions of intensity. This is reported to reduce testing time from days or weeks to hours. Ultraviolet radiation is so deleterious that if it is used care must be taken not to reduce test time to a few minutes, with consequent crowding of reaction times at the expense of differentiation of samples.

3. Room odour

When oils are heated to frying temperatures, odours may develop which are undesirable, although this is less noticeable with the newer ventilation systems in the cooking areas. Unfortunately, any off-odours are usually more noticeable to someone entering the area than to the person working there. For this reason, some customers prefer a "room odour" test similar to that developed by Evans *et al.*[1]* In this test, two rooms (8 × 5 × 10 ft), each provided with two small entry chambers to avoid premature exposure to odours are used. The chambers are at positive pressure relative to the odour rooms with all air exhausted at floor level. Oils (about 200 g) are heated to 190°C in glass dishes located in hoods exterior to the rooms. Odours are pumped into the rooms for organoleptic evaluation by trained judges.

D. FLAVOUR EVALUATION

Sensory evaluation panels are of two general types: affective (preference) evaluation and analytical (discriminative) evaluation. Consumer preference panels represent the affective type. They are made up of large numbers of untrained people who merely indicate their preference or acceptance. Quality control and research panels represent the discriminative type. They are made up of a few judges who have demonstrated flavour sensitivity and who have been trained by judges with long experience in the field of their particular product.

Organoleptic evaluation of oils is a difficult task with much opportunity for error, although under the proper conditions it yields very meaningful results. Trained tasters are essential. For example, many new technical employees of research organisations do not recognise the taste of so-called "light struck" soybean oil and grade it high. This may be due to their experience at home with oils which had been standing on supermarket shelves under fluorescent light for some time before being purchased. "Light struck" oil, however, will soon make itself noticed if used in salad dressing or mayonnaise.

1. Panel training

This is a complex subject and there are many literature references and books on the subject. Basically, volunteers are selected who have an interest in tasting and who have passed a preliminary screening test to see if they can detect flavour differences. Next comes the tasting of known score oil previously evaluated by expert tasters. After a panellist is "qualified" by performance on samples of known good and fair quality, he may participate in regular panels, keeping his score out of the evaluation until a comparison with other panellists demonstrates adequate training. As a necessary part of training, oil samples of varying quality should be obtained or prepared experimentally and scored by experts for use as "standards" in flavour evaluation. Each time oils are to be evaluated a "standard" sample of known quality should be included as a reference point. Even so-called expert tasters occasionally have an "off" day and the "standard" sample is of help in checking this.

* Superscript numbers refer to References at end of Chapter.

2. Uniform presentation

Samples should be presented in a standardised way as to quantity, container, and temperature. As little as 10 ml may be used, although most tasters prefer a somewhat larger quantity. The container should not introduce any foreign taste and, if lighting is not adjusted, the container should mask the colour of the sample. There is a tendency to downgrade darker oils and those with any unusual colour as, for example, those containing added carotene. Cold samples also tend to cover up flavour defects so oils should be warmed to around 43°C (110°F) before presentation. Experts still argue over the order of presentation of samples. Statistically, a random order of presentation seems desirable, but many experienced panellists prefer to smell the samples and try to taste the most bland samples first. The contention is that a very poor sample can affect the taste buds so that the flavour of later oils appears better or worse than it actually is. Rinsing the mouth with water, various solutions, or eating crackers are frequently used to clear the taste buds, but their efficacy is uncertain.

3. Scoring scale

In grading oils for quality, i.e. degree of rancidity or oxidative damage, a scale of 1–9 or 10 is commonly used. Usually, the higher the number the higher the quality, although some prefer to call "1" the top quality. It is most difficult to decide what degree of off-flavour each score number should represent, and this has to be done in consultation with tasters of long experience. Table 19.1 illustrates one possible scoring sheet. Using this form, the panellists decide the location of oil on a flavour intensity scale and then check some typical flavour characteristics detected which affect the score. For example, a moderate buttery flavour might be acceptable in a given sample, but slight "fishiness" would normally drop the grade precipitiously. In the last column, the overall acceptance is given, and this is the grade that counts. Samples grading 6 or better would normally be considered acceptable in quality, with 5 borderline, and below this unacceptable. The individual plant or laboratory must decide where they want their cut-off to be. Too much reliance cannot be placed on the flavour description, since any one oil, not completely bland, may be described by different panellists using a variety of terms. Terms, such as metallic, tallowy, burnt, musty, etc.,

TABLE 19.1. *Oil Evaluation Score Sheet*

Date
Judge
Sample Identification
Code

Flavour intensity	Flavour characteristics		Overall acceptance	
9—None	Bland	Raw	9	Acceptable range
8	Buttery	Oxidised	8	
7—Weak	Nutty	Rancid	7	
6	Beany	Fishy	6	
5—Moderate	Grassy	Painty	5	Marginal
4	Hydrogenated	Foreign	4	Unacceptable range
3—Strong	Light struck	Other	3	
2			2	
1—Extreme			1	

are often used, with little uniformity among judges. In addition, in training, a sample of real grass tastes "grassy" to practically everyone, but there is little agreement on what constitutes a "grassy" oil.

4. Statistical evaluation of data

In the testing just described the "mean" score of the panel is determined for each sample, along with the standard error or variance of the mean. This makes it possible to decide whether the score difference between two samples is greater than would be expected by chance alone. "Significant" normally means that the difference is greater than would be expected at a rate of 1 out of 20 trials. This is described as a probability of 0.05 or 5 %. "Very significant" means a chance probability of only 1 in 100 trials, 0.01 or 1 %. Statistical help should be obtained before setting up any oil evaluation programme.

E. ANTIOXIDANTS

Antioxidants may be defined as substances which, in small amounts, interfere with the normal oxidation process in oils and fats so as to delay the time when oxidation will have proceeded far enough to produce noticeable off-flavour and/or odours. They are usually thought to function as free radical acceptors, thus terminating the oxidation at the initiation step as shown in Fig. 19.2. In the absence of an antioxidant, a hydrogen atom is lost from the allylic carbon in the fatty acid group with the formation of a fatty free radical (R·). The latter is readily susceptible to attack by atmospheric oxygen, resulting in the formation of peroxides and hydroperoxides. When, for example, a phenolic-type antioxidant is present, it functions as a free radical acceptor forming a stable compound that will not propagate further oxidation of the glyceride. Here the antioxidant is a "primary" one. There is also a second class, so-called "synergists", which promote or "synergise" the action of other antioxidants, but having little effect if present alone. Examples are citric, phosphoric, thiodipropionic, ascorbic, and tartaric acids. These function by tying up ("chelating") pro-oxidant metals in the oil and to some extent by inhibiting peroxide decomposition and by regeneration or sparing of primary antioxidants.

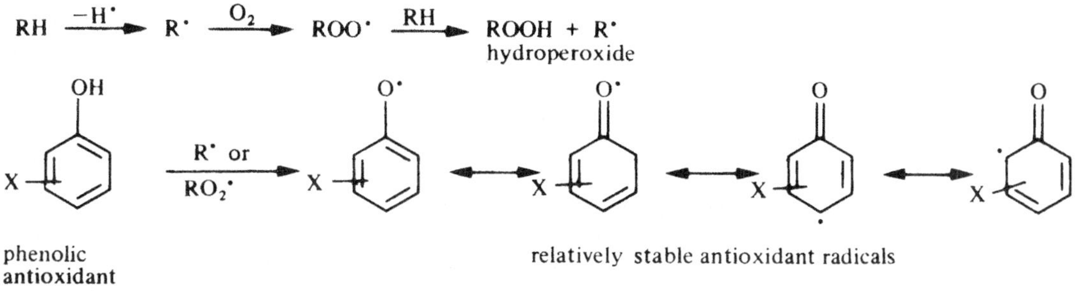

FIG. 19.2. Role of antioxidants.

It must be remembered in connection with the use of all antioxidants that each country has its own list of approved substances and permitted levels. These should be consulted in considering any proposed use. A review of world usage was published in 1978;[2] but in view of continuing changes as time goes on, it is necessary to make certain that the very latest regulations have been examined.

1. Primary antioxidants

(a) *Tocopherols.* Natural fats and oils are much more resistant to oxidation than pure triglycerides because of the presence of antioxidants, so-called "inhibitols", in the former. Moreover, vegetable oils resist oxidative rancidity better than animal fats because of their higher content of these natural antioxidants, now identified mostly as tocopherols. The antioxidant action of tocopherols was first demonstrated by Olcott and Emerson[3], and later tocopherols were identified as the active substances in the "inhibitols" previously isolated from a variety of vegetable oils.[4]

The four principal tocopherols are designated alpha, beta, gamma, and delta (α, β, γ, δ). These tocopherols (**1**) differ in the number and location of methyl groups in the aromatic ring. In general, antioxidant activity diminishes as one goes from δ to γ to β to α tocopherol.

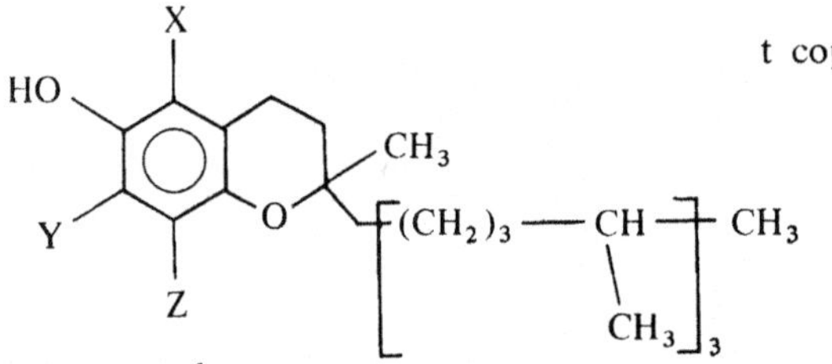

1

t copherol	*X*	*Y*	*Z*
α	CH_3	CH_3	CH_3
β	CH_3	H	CH_3
γ	H	CH_3	CH_3
δ	H	H	CH_3

A representative soybean oil will contain α-tocopherol (0.009–0.12 %), γ-tocopherol (0.074–0.102 %), and δ-tocopherol (0.024–0.030 %), for a total of 0.107–0.144 %. Refining will reduce these to roughly half their content in crude oil. β-Tocopherol makes up less than 3 % of total tocopherol content.

It is commonly agreed that too much as well as too little tocopherol is undesirable. In general, levels between 0.03 and 0.10 % appear optimum. Tocopherols are most effective in the absence of light and in animal fats lacking natural antioxidants.

(b) *Gum guaiac.* This is a resinous exudate from a type of tree found in the West Indies. It is phenolic in nature. Cleared in the United States in 1940 for animal fat applications, it was for some time the only antioxidant available for that purpose. It has little efficacy in vegetable oils and is subject to colour and flavour problems.

(c) *Gallic acid esters.* Several alkyl gallates were recommended as antioxidants in the early 1940s and were shortly authorised for use in a number of countries. The higher esters (octyl and dodecyl) are more fat-soluble than the lower esters, but all gallates tend to show blue-black discolouration in the presence of traces of iron. Gallates are also heat-sensitive, especially under alkaline conditions, and tend to be lost rapidly from cooking oils at high temperatures.

Table 19.2 includes some stability data on gallates.

(d) *Nordihydroguiaretic acid* (*NDGA*). Nordihydroguiaretic acid (**2**), extracted from the creosote bush, was cleared as an antioxidant for meat fats in the United States in the early 1940s. It possesses phenolic properties like the gallates, with the same advantages and disadvantages. Some data are shown in Table 19.2. High cost has contributed to its limited use.

TABLE 19.2. *Typical Antioxidant Effectiveness of gallates, NDGA and TBHQ in Vegetable Oils*[5]

	A.O.M. stability (hours to 70 P.V.)		
% Antioxidant in oil	Cottonseed	Soybean	Palm
Propyl gallate			
0.01	19(9)*	21(11)	137(45)
0.02	30(9)	26(11)	—
NDGA			
0.02	14(12)		—
TBHQ			
0.01	24(9)	29(11)	150(45)
0.02	34(9)	41(11)	

* Numbers in parenthesis refer to values for oil without antioxidant.

(e) *Butylated hydroxy anisole (BHA)*. Butylated hydroxyanisole is a mixture of **3a** (~15%) and **3b** (~85%). It was approved in the United States in 1948 for food use and has since found widespread use in other countries. Although not very effective in vegetable oils. BHA antioxidant action has a carry-through effect in baked and fried foods containing fats and oils. It may be used in combination with other primary antioxidants. It possesses a noticeable phenolic odour, especially when heated to high temperatures.

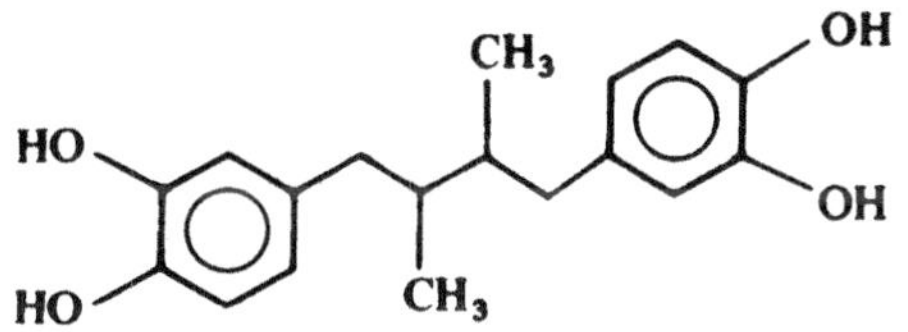

2

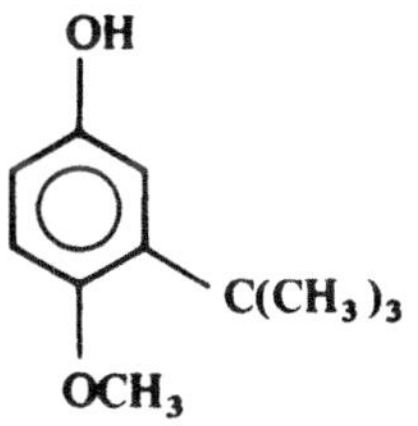

3a

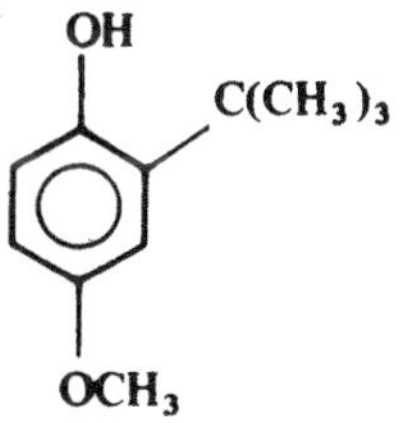

3b

(f) *Butylated hydroxytoluene (BHT)* Butylated hydroxytoluene (**4**), previously used extensively in non-food applications, was cleared in the United States in 1954 for use in food oils. It has relatively low antioxidant effectiveness in vegetable oils, but, like BHA, is often used with other primary antioxidants because of its carry-over effectiveness in baked or fried products. Pending tests in progress in connection with a review of all "GRAS" (Generally Regarded As Safe) substances, the use of BHT is temporarily restricted in the United States to current levels in foods for which it is now approved.

(g) *Tertiary butyl hydroquinone (TBHQ)* Tertiary butyl hydroquinone (**5**) was cleared in the United States in 1972. As seen in Table 19.2, it is more effective in vegetable oils than any of the other primary antioxidants. It also appears to surmount many of the colour and solubility problems encountered with other antioxidants.

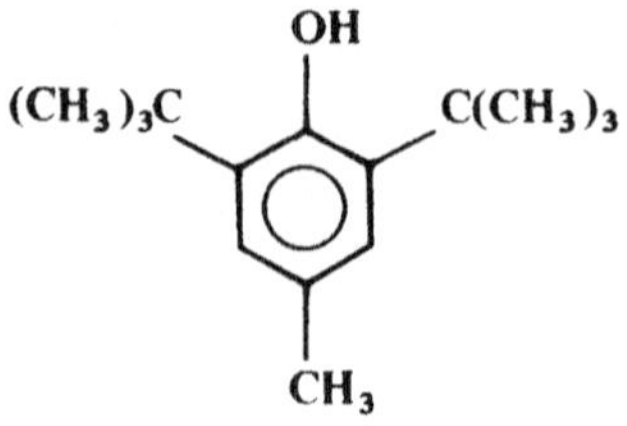

4

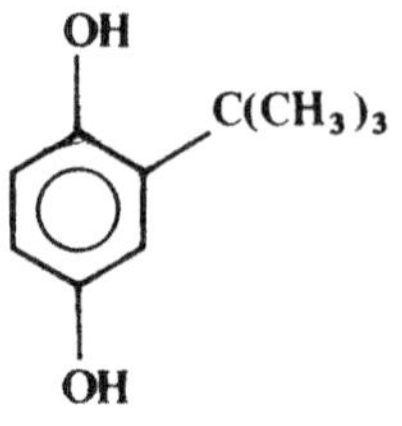

5

2. Synergistic antioxidants

Although the exact nature of the synergism is not well understood, much of the effect may be due to inactivation of prooxidant metals present in the oil as a result of chelation with the metals. Table 19.3 shows the concentration of various metals required to reduce the keeping time of lard by 50 % when it is held at 98°C.[6] Copper and iron are most important on a practical basis. It is often assumed that these metals are present as soaps formed by the action of free fatty acids in the oil during processing from seed to crude oil; however, this could not be demonstrated in several plant tests conducted by the writer. It appears more likely that the metals are bound organically in some form other than soap as a prooxidant component of the seed. This must be broken down and the metal inactivated by chelation with such synergists as citric acid. Some heat is generally required.

TABLE 19.3. *Metal Concentrations Necessary to Reduce the Keeping Time of Lard by 50 % at 98°C*[6]

Metal	Concentration in ppm
Copper	0.05
Manganese	0.6
Iron	0.6
Chromium	1.2
Nickel	2.2
Vanadium	3.0
Zinc	19.6
Aluminium	50.0

Unfortunately, present metal analyses do not permit the measurement of chelated metals (as with citric acid) separately from total metal content. Thus, the only option is to try to reduce total metal content (mainly iron and copper) to as low a level as possible. Evidence of chelation is essentially indirect, based mainly on A.O.M.s of oils with and without treatment with citric acid or the like. When prooxidant metals are inactivated or removed, primary antioxidants have an easier task. On the other hand, in the presence of prooxidant metals, at least in vegetable oils like soybean, primary antioxidants are of little value. It is, therefore, recommended that citric acid always be added in the cooling stage of deodorisation. Since citric acid decomposes at 150°C (302°F), well below deodorisation temperatures, this practice appears logical, although, for some reason, a few processors have had good results adding citric before deodorisation.

Although citric acid is the most commonly used synergist (or metal chelant), other acids, like ascorbic and tartaric, have been used. The phospholipid, lecithin, is also effective, but may introduce colour or flavour problems. Phosphoric acid is very effective, but the level of usage is critical. Excessive amounts can cause the development of melon-like and cucumber-like off-flavours with resultant lower flavour scores in aged oils, despite improved oxidative stability.

REFERENCES

1. C. D. Evans, H. A. Moser, G. R. List, H. J. Dutton, and J. C. Cowan, *J. Amer. Oil Chemists' Soc.* 1971, **48,** 840.
2. Y. Botma, *Food Engineering*, 93 (May 1978).
3. H. S. Olcott and O. H. Emerson, *J. Amer. Chem. Soc.* 1937, **59,** 1008.
4. H. S. Olcott and H. A. Mattill, *J. Amer. Chem. Soc.* 1936, **58,** 1627.
5. E. R. Sherwin, *J. Amer. Oil Chemists' Soc.* 1976, **53,** 430.
6. R. Ohlson, "Fats and Oils Demetalization. Its Influence on Their Oxidative Stability", *Proceedings of the Third International Symposium on Metal-Catalyzed Lipid Oxidation*, Paris, Institute de Corps Gras, Sept. 27–30 (1973).

GENERAL REFERENCES

Buck, D. F., "Antioxidants in Soybean Oil", *J. Amer. Oil Chemists' Soc.* 1981, **58,** 275.

Erickson, D. R., "Finished Oil Handling and Storage", *J. Amer. Oil Chemists' Soc.* 1978, **55,** 815.

Erickson, D. R., E. H,. Pryde, O. L. Brekke, T. L. Mounts and R. A. Falb, (eds.) *Handbook of Soy Oil Processing and Utilization.* American Soybean Association and the American Oil Chemists' Society, 1980.

Hall Ellis, B., "Acceptance and Consumer Preference Testing", *J. Dairy Science*, 1969, **52,** 823.

Jackson, H. W., "Techniques for Flavour and Oil Evaluation", *J. Amer. Oil Chemists' Soc.* 1981, **58,** 227.

Larmond, E., *Methods for Sensory Evaluation of Food.* Publication 1284, Canada Department of Agriculture, 1967.

Sherwin, E. R., "Antioxidants for Vegetable Oils", *J. Amer. Oil Chemists' Soc.* 1976, **53,** 430; *ibid*, 1978, **55,** 809.

Index

Printed in Dunstable, United Kingdom

91221469R00105